Methods in Molecular Biology

Series Editor
John M. Walker
School of Life and Medical Sciences
University of Hertfordshire
Hatfield, Hertfordshire, AL10 9AB, UK

For further volumes:
http://www.springer.com/series/7651

Brassinosteroids

Methods and Protocols

Edited by

Eugenia Russinova

VIB-Ghent University, Departments of Plant Systems Biology and Plant Biotechnology and Bioinformatics, Ghent, Belgium

Ana I. Caño-Delgado

Research in Agricultural Genomics (CRAG), CSIC-IRTA-UAB-UB, Cerdanyola del Vallès, Barcelona, Spain

Humana Press

Editors
Eugenia Russinova
VIB-Ghent University
Departments of Plant Systems Biology
and Plant Biotechnology and Bioinformatics
Ghent, Belgium

Ana I. Caño-Delgado
Research in Argicultural Genomics (CRAG)
CSIC-IRTA-UAB-UB
Cerdanyola del Vallès, Barcelona, Spain

ISSN 1064-3745 ISSN 1940-6029 (electronic)
Methods in Molecular Biology
ISBN 978-1-4939-6811-4 ISBN 978-1-4939-6813-8 (eBook)
DOI 10.1007/978-1-4939-6813-8

Library of Congress Control Number: 2016961692

Cover Illustration: The book cover image illustrates the localization of brassinosteroid receptor BRL3 enriched at stem cell niche of the root meristem in Arabidopsis. Understanding the molecular bases for cell-specific action of brassinosteroid signalling opens new and exciting possibilities for understanding plant growth and adaption to the changing environment. Image taken with permission from ASPB as reported in: Fabregas, N, Li, N, Boeren, S., Nash, T.E., Goshe, M.B., Clouse S.D., de Vries S. and Caño-Delgado, A.I. Specific roles for BRL3 signalosome complex in Arabidopsis root development. Plant Cell, 25(9); 3377-3388 (2013). doi: 10.1105/tpc.113.114462

Printed on acid-free paper

This Humana Press imprint is published by Springer Nature
The registered company is Springer Science+Business Media LLC
The registered company address is: 233 Spring Street, New York, NY 10013, U.S.A.

Preface

Brassinosteroids are naturally occurring endogenous plant hormones indispensable for growth and development. Since their discovery, intense research based on genetic analysis of brassinosteroid-deficient and brassinosteroid-insensitive mutants, and on proteomics and genomics mainly in the model plant species *Arabidopsis thaliana* and *Oryza sativa* (rice) have yielded a comprehensive knowledge on the brassinosteroid signal transduction pathway, from receptor kinases to transcription factors and their targets, so that it has become one of the best characterized plant hormone pathways. Although originally described as linear, the brassinosteroid signal transduction pathway is highly interconnected with other signaling pathways, including auxin, gibberellic acid, abscisic acid, light, and sugar, thus supporting the notion that brassinosteroids are master regulators of plant growth. The brassinosteroid signal transduction pathway also shares components of and interacts with other receptor kinase pathways that regulate immunity and stomatal, root, and reproductive development. The prospective agronomical importance of the brassinosteroid hormones requires active research in plant species other than *Arabidopsis*, but preliminary results hint at a good level of transferability between dicotyledonous and monocotyledonous plant species, regarding the effects of brassinosteroids on plant architectural traits. Therefore, a potential exists to identify brassinosteroid mutants or transgenic plants with improved productivity in agronomically important crops, such as maize (*Zea mays*), wheat (*Triticum* sp.), sorghum (*Sorghum bicolor*), soybean (*Glycine max*), potato (*Solanum tuberosum*), poplar (*Populus* sp.), and tomato (*Solanum lycopersicum*).

Despite the important advances made in the brassinosteroid field, a number of essential questions remain to be answered. Research focused on resolving specific brassinosteroid signaling pathways for different cellular lineages has just started to develop. The advent of single-cell genomics technologies stands as a promise for the rapid identification of brassinosteroid pathways with single-cell resolution. Whereas the long-distance transport of brassinosteroid hormones has been ruled out, the mechanisms for brassinosteroid short-range movement remain a mystery. How brassinosteroid molecules are secreted from the cell where they are synthesized also needs to be clarified, because it is a prerequisite to elucidate the cell-to-cell signaling governed by the hormones.

In this book, we compiled state-of-the-art methodologies for the study of brassinosteroid hormones. In 16 chapters, recognized researchers in the brassinosteroid field bring together different experimental and theoretical biology methods for addressing the questions of how brassinosteroids function in *Arabidopsis* and other agriculturally valuable species, such as rice and sorghum. The topics cover a wide range of protocols to analyze brassinosteroid levels and synthesize fluorescently labeled brassinosteroid analogs for monitoring brassinosteroids in living cells (Chapters 1 and 2). We present detailed procedures for physiological analysis of brassinosteroid responses in *Arabidopsis* and in rice to study the crosstalk with other signaling pathways, such as light and immunity (Chapters 3, 4, 5, and 6). Chapters 7 and 8 describe genomic methods for the identification of brassinosteroid-regulated genes and networks in the whole plant or in specific tissues. This book also contains mathematical modeling approaches to study the dynamics of brassinosteroid signaling components (Chapter 9) and quantitative microscopy methods to monitor the relative levels of the BRASSINOSTEROID INSENSITIVE1 (BRI1) protein in the plasma membrane (Chapter 10). We included biochemical and mass spectrometry protocols for purification and identification of brassinosteroid signaling components (Chapters 11 and 12) and for the analysis of the posttranslational modifications of the BRI1 receptor (Chapters 13 and 14). Furthermore, the potential of brassinosteroid molecules in agriculture is shown in Chapter 15, in which sorghum is proposed as a model crop species for the study of brassinosteroid signaling in root growth and development. Protocols to assess brassinosteroid effects on abiotic stress tolerance in *Arabidopsis* and *Brassica napus* (rapeseed) are presented in Chapter 16. We hope this book will not only serve as an ideal reference for researchers in the brassinosteroid field but also assess a wide range of appealing methodologies for researchers with a general interest in hormone signaling in plants. We thank all our colleagues who contributed to this book and Martine De Cock for help in editing the manuscripts.

Gent, Belgium — *Eugenia Russinova*
Barcelona, Spain — *Ana I. Caño-Delgado*

Contents

Contributors

ELIZABETH BARTOM • *Department of Biochemistry and Molecular Genetics, Northwestern University Feinberg School of Medicine, Chicago, IL, USA*

JOS VAN DEN BEGIN • *Organic and Biomimetic Chemistry Research Group, Department of Organic and Macromolecular Chemistry, Ghent University, Gent, Belgium*

YOUSSEF BELKHADIR • *Gregor Mendel Institute (GMI), Vienna BioCenter (VBC), Austrian Academy of Sciences, Vienna, Austria*

DAVID BLASCO-ESCÁMEZ • *Department of Molecular Genetics, Center for Research in Agricultural Genomics (CRAG), Consejo Superior de Investigaciones Científicas-Institut Recerca i Tecnologia Agroalimentàries-Universitat Autonóma de Barcelona-Universitat de Barcelona (CSIC-IRTA-UAB-UB), Barcelona, Spain*

SJEF BOEREN • *Laboratory of Biochemistry, Wageningen UR, Wageningen, The Netherlands*

ANA I. CAÑO-DELGADO • *Department of Molecular Genetics, Center for Research in Agricultural Genomics (CRAG), Consejo Superior de Investigaciones Científicas-Institut Recerca i Tecnologia Agroalimentàries-Universitat Autonóma de Barcelona-Universitat de Barcelona (CSIC-IRTA-UAB-UB), Barcelona, Spain*

SUNGHWA CHOE • *School of Biological Sciences, College of Natural Sciences, Seoul National University, Seoul, Korea; Convergence Research Center for Functional Plant Products, Advanced Institutes of Convergence Technology, Gyeonggi-do, Korea; Plant Genomics and Breeding Institute, Seoul National University, Seoul, Korea*

CHENGCAI CHU • *State Key Laboratory of Plant Genomics and National Center for Plant Gene Research (Beijing), Institute of Genetics and Developmental Biology, Chinese Academy of Sciences, Beijing, China*

CARINA A. COLLINS • *Division of Biochemistry; Interdisciplinary Plant Group (IPG), University of Missouri, Columbia, MO, USA*

CLAUDIA CORVALÁN • *School of Biological Sciences, College of Natural Sciences, Seoul National University, Seoul, Korea*

WALTER VAN DONGEN • *Laboratory of Biochemistry, Wageningen UR, Wageningen, The Netherlands*

NORMA FÀBREGAS • *Department of Molecular Genetics, Center for Research in Agricultural Genomics (CRAG), Consego Superior de Investigaciones Científicas-Institut Ricerca i Tecnologia Agroalimentaries-Universitat Autonóma de Barcelona-Universitat de Barcelona (CSIC-IRTA-UAB-UB), Barcelona, Spain*

DAVID FRIGOLA • *Departament de Física de la Matèria Condensada, Facultat de Física, Universitat de Barcelona, Barcelona, Spain*

HONGQING GUO • *Department of Genetics, Development and Cell Biology, Iowa State University, Ames, IA, USA*

LUC VAN HEERDE • *Laboratory of Biochemistry, Wageningen UR, Wageningen, The Netherlands*

ANTJE HEESE • *Division of Biochemistry; Interdisciplinary Plant Group (IPG), University of Missouri, Columbia, MO, USA*

MARTA IBAÑES • *Departament de Física de la Matèria Condensada, Facultat de Física, Universitat de Barcelona, Barcelona, Spain*

NILOUFER G. IRANI • *Department of Plant Systems Biology, VIB, Gent, Belgium; Department of Plant Biotechnology and Bioinformatics, Ghent University, Gent, Belgium; Department of Plant Sciences, University of Oxford, Oxford, UK*

YVON JAILLAIS • *Laboratoire de Reproduction et Développement des Plantes, Université de Lyon, Ecole Nationale Supérieure de Lyon, Université Claude Bernard Lyon 1, Centre National de la Recherche Scientifique, Institut National de la Recherche Agronomique, Lyon, France*

PRITI KRISHNA • *School of Environmental and Rural Sciences, University of New England, Armidale, NSW, Australia*

MICHELLE E. LESLIE • *Division of Biochemistry; Interdisciplinary Plant Group (IPG), University of Missouri, Columbia, MO, USA*

LEI LI • *Department of Genetics, Development and Cell Biology, Iowa State University, Ames, IA, USA; Harvard Medical School, Harvard University, Boston, MA, USA*

SANZHEN LIU • *Department of Plant Pathology, Kansas State University, Manhattan, KS, USA*

ROSA LOZANO-DURÁN • *Shanghai Center for Plant Stress Biology (PSC), Shanghai Institutes for Biological Sciences (SIBS), Chinese Academy of Sciences (CAS), Shanghai, China*

FIDEL LOZANO-ELENA • *Department of Molecular Genetics, Center for Research in Agricultural Genomics (CRAG), Consejo Superior de Investigaciones Científicas-Institut Recerca i Tecnologia Agroalimentàries-Universitat Autonóma de Barcelona-Universitat de Barcelona (CSIC-IRTA-UAB-UB), Barcelona, Spain*

YU LUO • *Department of Plant Systems Biology, VIB, Gent, Belgium; Department of Plant Biotechnology and Bioinformatics, Ghent University, Gent, Belgium*

ANNEMIEKE MADDER • *Organic and Biomimetic Chemistry Research Group, Department of Organic and Macromolecular Chemistry, Ghent University, Gent, Belgium*

JAIME F. MARTÍNEZ-GARCÍA • *Centre for Research in Agricultural Genomics (CRAG), Consego Superior de Investigaciones Científicas-Institut de Recerca i Tecnologia Agroalimentaries-Universitat Autonóma de Barcelona-Universitat de Barcelona (CSIC-IRTA-UAB-UB), Barcelona, Spain; Institució Catalana de Recerca i Estudis Avançats (ICREA), Barcelona, Spain*

SARA MARTINS • *Institute for Integrative Biology of the Cell (I2BC), Centre National de la Recherche Scientifique/Commissariat à l'Energie Atomique/Université Paris Sud, Université Paris-Saclay, Gif-sur-Yvette, France*

MARIA JOSÉ MOLINA-CONTRERAS • *Centre for Research in Agricultural Genomics (CRAG), Consego Superior de Investigaciones Científicas-Institut de Recerca i Tecnologia Agroalimentàries-Universitat Autonòma de Barcelona-Universitat de Barcelona (CSIC-IRTA-UAB-UB), Barcelona, Spain*

TREVOR NOLAN • *Department of Genetics, Development and Cell Biology, Iowa State University, Ames, IA, USA*

SANDI PAULIŠIĆ • *Centre for Research in Agricultural Genomics (CRAG), Consego Superior de Investigaciones Científicas-Institut de Recerca i Tecnologia Agroalimentàries-Universitat Autonóma de Barcelona-Universitat de Barcelona (CSIC-IRTA-UAB-UB), Barcelona, Spain*

SCOTT C. PECK • *Division of Biochemistry; Interdisciplinary Plant Group (IPG), University of Missouri, Columbia, MO, USA*

BRIGITTE POPPENBERGER • *Biotechnology of Horticultural Crops, TUM School for Life Sciences Weihenstephan, Technical University Munich, Freising, Germany*

BISHUN D. PRASAD • *Department of Molecular Biology and Genetic Engineering, Bihar Agricultural University, Sabour, Bihar, India*
TAWHIDUR RAHMAN • *Saskatoon Research Centre, Agriculture and Agri-Food Canada, Saskatoon, SK, Canada*
IRMA ROIG-VILLANOVA • *Centre for Research in Agricultural Genomics (CRAG), Consego Superior de Investigaciones Científicas-Institut de Recerca i Tecnologia Agroalimentàries-Universitat Autonòma de Barcelona-Universitat de Barcelona (CSIC-IRTA-UAB-UB), Barcelona, Spain*
WILFRIED ROZHON • *Biotechnology of Horticultural Crops, TUM School for Life Sciences Weihenstephan, Technical University Munich, Freising, Germany*
EUGENIA RUSSINOVA • *Department of Plant Systems Biology, VIB, Gent, Belgium; Departments of Plant Systems Biology and Plant Biotechnology and Bioinformatics, VIB-Ghent University, Gent, Belgium*
SIGAL SAVALDI-GOLDSTEIN • *Faculty of Biology, Technion-Israel Institute of Technology, Haifa, Israel*
PATRICK SCHNABLE • *Department of Agronomy, Ames, IA, USA*
MIROSLAV STRNAD • *Laboratory of Growth Regulators, Centre of the Region Haná for Biotechnological and Agricultural Research, Institute of Experimental Botany, Academy of Sciences of the Czech Republic, and Palacký University, Olomouc, Czech Republic*
DANUŠE TARKOWSKÁ • *Laboratory of Growth Regulators, Centre of the Region Haná for Biotechnological and Agricultural Research, Institute of Experimental Botany, Academy of Sciences of the Czech Republic, and Palacký University, Olomouc, Czech Republic*
HONGNING TONG • *State Key Laboratory of Plant Genomics and National Center for Plant Gene Research (Beijing), Institute of Genetics and Developmental Biology, Chinese Academy of Sciences, Beijing, China*
SIMON J. UNTERHOLZNER • *Biotechnology of Horticultural Crops, TUM School for Life Sciences Weihenstephan, Technical University Munich, Freising, Germany*
GRÉGORY VERT • *Institute for Integrative Biology of the Cell (I2BC), Centre National de la Recherche Scientifique, Commissariat à l'Energie Atomique, Université Paris Sud, Université Paris-Saclay, Gif-sur-Yvette, France*
KRISTINA VRAGOVIĆ • *Faculty of Biology, Technion-Israel Institute of Technology, Haifa, Israel*
SACCO C. DE VRIES • *Laboratory of Biochemistry, Wageningen UR, Wageningen, The Netherlands*
JOHAN M. WINNE • *Laboratory for Organic Synthesis, Department of Organic and Macromolecular Chemistry, Ghent University, Gent, Belgium*
YANHAI YIN • *Department of Genetics, Development and Cell Biology, Iowa State University, Ames, IA, USA*

Chapter 1

Protocol for Extraction and Isolation of Brassinosteroids from Plant Tissues

Danuše Tarkowská and Miroslav Strnad

Abstract

Brassinosteroids (BRs) are a class of steroid plant hormones that participate with other plant hormones in the regulation of numerous developmental processes, including root and shoot growth, vascular differentiation, fertility, and seed germination. A characteristic feature of all plant hormones, including BRs, is that their concentration is extremely low in plant tissues and, therefore, the methods dealing with their determination belong to ultra-trace analysis, for which very sensitive analytical tools are needed. The analysis of natural BRs is essential when their functions and roles in plant growth and development are to be elucidated. Here, we describe a reliable protocol for high-throughput extraction and purification of BRs. The procedure consists of two solid-phase extraction steps and provides selective enrichment and efficient cleanup of these compounds from complex plant extracts. The protocol is designed for sensitive liquid chromatography-tandem mass spectrometry-based method for simultaneous detection of 22 naturally occurring BRs, including their biosynthetic precursors and most of their biologically active metabolites, without need for derivatization.

Key words Brassinosteroids, Plant hormones, Solid-phase extraction, Ultra-trace analysis, Gas chromatography, Liquid chromatography, Mass spectrometry

1 Introduction

Brassinosteroids (BRs) constitute one of the eight known families of plant hormones, namely chemical messengers that play pivotal roles in cell–cell communication. Plants, being multicellular organisms, use these signaling molecules to coordinate and regulate growth and development of their cells, tissues, and organs as well as to respond to environmental stimuli. The plant steroid hormone group comprises approximately 70 naturally occurring substances that are, in terms of their core structure, related to insect and animal steroid hormones, i.e., they possess a tetracyclic (four-ringed) skeleton. Natural BRs have a skeleton derived from 5α-cholestane and contain 27 carbon atoms (Fig. 1). During BR biosynthesis in plants, rings A and B, as well as the cholestane side chain, are preferentially modified by selective hydroxylation that in the case of the

Eugenia Russinova and Ana I. Caño-Delgado (eds.), *Brassinosteroids: Methods and Protocols*, Methods in Molecular Biology, vol. 1564, DOI 10.1007/978-1-4939-6813-8_1, © Springer Science+Business Media LLC 2017

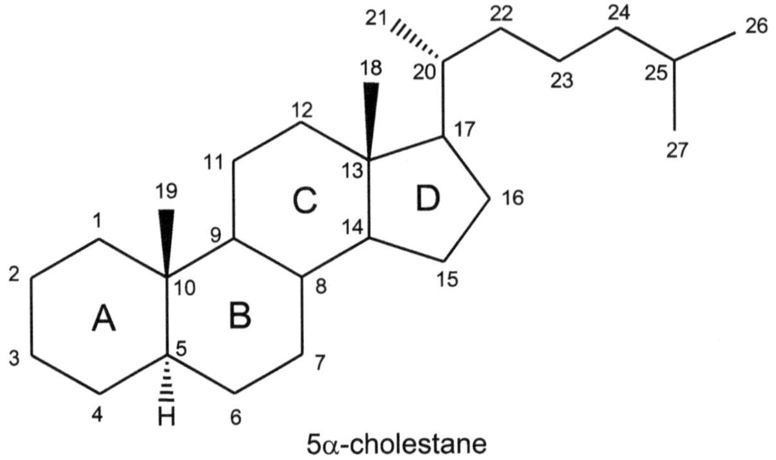

Fig. 1 Structure of 5α-cholestane

side chain takes place mainly at positions C-22 and/or C-23. The cholestane side chain can also undergo various oxidation or reduction reactions during the biosynthesis of plant sterols, leading to the addition or loss of different substituents at positions C-23, C-24, and/or C-25 [1].

As a result of these modifications, BRs can be classified as C_{27}, C_{28}, or C_{29} BRs, according to the number of carbon substituents on the C-24 of the side chain. Among these BR classes, the C_{28} BRs are the most omnipresent in nature, including, for example, castasterone (CS) and brassinolide (BL). Because of the chemical nature of plant steroidal hormones, as described above, these signaling molecules are highly hydrophobic substances compared to other plant hormones. This is the first important feature that should be taken into account when developing analytical methods for the determination of BRs in plant tissues. The second major factor influencing BR analysis is the extremely low concentrations (fg to pg/mg fresh weight) at which they are present and act in plant tissues of interest. For this reason, the analysis of BRs can be classified as ultra-trace analysis that requires very sensitive analytical tools. Undoubtedly, the analysis of plant BRs is essential to elucidate their functions and roles in plant growth and development. Although BRs have been detected in all plant organs, their concentrations differ depending on the type of tissue. In general, higher levels of BRs can be observed in reproductive tissues than in vegetative ones, the difference usually being three orders of magnitude [2]. A smaller difference in BR abundance can be found in young vs. mature tissue from the same plant organ. BRs are also abundant in crown gall and insect gall cells where their levels are higher than in normal cells [1]. It is important to stress here that plants are able to synthesize a great variety of sterols and that BRs represent only a minority of this huge group. Nevertheless, it is also essential to note that BRs are the only plant steroids with a hormonal activity mediated via a known signaling receptor pathway. Accordingly, the concentrations of BRs are much lower than those of other plant sterols and these extremely low concentrations are also unique within the family of plant hormones. Furthermore, concentrations of other substances originating from plant tissue matrices that interfere with the BR analysis are far higher than those of the BRs themselves. This fact is the major problem associated with plant hormone analysis, in general. Thus, to accurately quantify plant hormones, it is of great importance not only that the chemical characteristics of the hormones to be analyzed are well understood, but also that the identities of potentially interfering substances are known.

Sound knowledge of the chemical principles regarding the extraction and purification of BRs and all the other substances present in plant tissues is also indispensable for accurate and precise hormone determination. In general, the solvent used must be capable of extracting the hormone efficiently, while minimizing the extraction of interfering substances. BRs as neutral compounds that display no ionic properties and a high hydrophobicity are most often extracted in organic solvents, such as methanol or acetonitrile [3–6]. Usually, prior to the analysis based on gas chromatography, eight purification steps have to be performed, making the sample preparation extremely labor intensive and time consuming. These steps are: two partitioning steps, two column chromatography steps (a charcoal column followed by a Sephadex LH-20 column), two solid-phase extraction steps (with diethylaminopropyl silica and octadecyl silica sorbents), high performance liquid chromatography (HPLC) fractionation, and finally a derivatization step. Excepting methodologies based on these demanding procedures, a few methods have been published for easier plant extract purification with maintenance of purification efficiency and degree. These solid-phase extraction-based approaches often precede the analysis of BRs by liquid chromatography [6–8].

We have recently developed a sensitive mass spectrometry-based method [9] for the simultaneous profiling of 22 naturally occurring BRs, including their biosynthetic precursors and the majority of their biologically active metabolites, without derivatization. Using ultra-high performance liquid chromatographic (UHPLC) separation, BRs were detected in the highly selective multiple reaction monitoring mode. The detection limit for most of the BRs analyzed was close to 2 pg. These low detection limits for a broad range of BR metabolites made it possible for as little as 50 mg of plant tissue to be used for quantitative analyses. The method was optimized for *Brassica napus* (oilseed rape) flowers and *Arabidopsis thaliana* plant tissue extracts purified by two-step solid-phase extraction (SPE). Determination with internal standards showed that this approach provides a high degree of feasibility, reproducibility, and recovery. Recent results with various plant materials [10–13] demonstrate the usefulness of the proposed SPE/UHPLC–MS/MS protocol for the estimation of natural BRs. Analysis of tissues originating from different plant species revealed that not all plants contain all BRs: CS is the most widely distributed, whereas BL, the most biologically active BR, occurs much less frequently. Interestingly, there are plant species that contain CS and BL without their C-24 stereoisomers *epi*BL and *epi*CS, differing in configuration at one stereocenter in the BR molecule, at any detectable level; these species include *Arabidopsis*, bean (*Phaseolus* sp.), and oilseed rape [1, 3]. Furthermore, BL and CS are often found in plants belonging to the Brassicaceae family in the presence of their counterparts with one carbon less (*nor*BL and *nor*CS). Among the C_{29} BRs, homoCS is often detected in most plant species. The distribution of the direct biosynthetic precursor of CS typhasterol (TY) is usually similar to that of CS itself, whereas the occurrence of teasterone, the precursor of TY, is less frequent.

2 Materials

2.1 Solutions

1. 60 % (v/v) acetonitrile (ACN; gradient grade for liquid chromatography) in deionized water.
2. 100 % (v/v) methanol (MeOH; gradient grade for liquid chromatography).
3. 10 % (v/v) MeOH (gradient grade for liquid chromatography) in deionized water (*see* **Note 1**).
4. Freely selectable stable isotope-labeled internal standards of BRs dissolved in MeOH or dimethyl sulfoxide.

2.2 Plant Material

Arabidopsis thaliana (L.) Heynh., *Brassica napus*, or other plant species belonging to *Brassicaceae* family.

2.3 Equipment and Other Materials

1. Ball mill.
2. Ceria (cerium oxide)-stabilized zirconium oxide balls.
3. Laboratory rotator.
4. General purpose centrifuge that can reach 36,670 × *g*.
5. Solid-phase extraction accessories and vacuum manifold.
6. Discovery DPA-6S cartridges (50 mg bed size).
7. Isolute C4 SPE cartridges (100 mg bed size).
8. Benchtop concentrator for evaporation under vacuum.
9. Vortex.
10. Ultrasonic bath.

11. Mortar and pestle.
12. 12 × 75 mm disposable borosilicate glass tubes.
13. 2-mL Eppendorf tubes.

3 Methods

3.1 Extraction

1. Grind fresh plant tissue frozen in liquid nitrogen in a mortar with a pestle.
2. Weigh 50 mg of homogenized plant tissue into 2-mL Eppendorf tubes, add 1 mL of 60 % (v/v) ACN, 2 to 3 zirconium balls, and an appropriate amount of stable isotope-labeled internal BRs standards (*see* **Note 2**).
3. Homogenize the sample with a ball mill at a frequency of 30 Hz for 3 min.
4. Sonicate the tubes with the crude extract for 5 min and extract overnight with a laboratory rotator at 17 rpm and 4 °C.
5. Centrifuge the samples at 36,670 × *g* and 4 °C for 10 min; transfer the supernatants into clean borosilicate glass test tubes.

3.2 Purification Using Discovery DPA-6S cartridges

1. Activate the sorbent in a cartridge with 1 mL 100 % (v/v) MeOH.
2. Equilibrate the sorbent in a cartridge with 1 mL 60 % (v/v) ACN.
3. Load 1 mL of sample (supernatant) in 60 % (v/v) ACN (*see* Subheading 3.1, **step 5**) at a flow rate of 0.75 mL/min. Collect flow-through fraction containing BRs into clean borosilicate glass test tubes.
4. Evaporate the fraction in a test tube to dryness with a benchtop concentrator.
5. Dissolve the sample residue in a test tube in 100 μL of ice-cold 100 % (v/v) MeOH; sonicate for 5 min, vortex, add 900 μL deionized water (*see* **Note 1**), and purify through Isolute® C4 cartridges.

3.3 Purification Using Isolute C4 Cartridges

1. Activate the sorbent in a cartridge with 1 mL 100 % (v/v) MeOH.
2. Equilibrate the sorbent in a cartridge with 1 mL 10 % (v/v) MeOH.
3. Load 1 mL of sample purified through Discovery DPA-6S cartridges and dissolved as mentioned above (*see* Subheading 3.2, **step 5**) at a flow rate of 1–2 mL/min.
4. Wash the sorbent in a cartridge with 1 mL 10 % (v/v) MeOH.

5. Elute BRs from the sorbent in a cartridge with 1 mL of 100 % (v/v) MeOH.
6. Evaporate the eluate to dryness under vacuum.
7. Dissolve the sample in a mobile phase or a solvent appropriate for the analytical instrument chosen for the quantitative analysis of BRs.

4 Notes

1. Deionized water means Milli-Q water, i.e., water prepared by Simplicity 185 or higher version of the Millipore water system.
2. The amount of internal standard depends on sensitivity of the analytical instrument used. Each experimenter has to optimize it according to his instrumental equipment available.

Acknowledgments

The authors thank Sees-editing Ltd. for critical reading and editing of the manuscript. The work was supported by the Grant Agency of the Czech Republic (14-34792S) and by the Ministry of Education, Youth and Sport of the Czech Republic through National Program of Sustainability (LO 1204).

References

1. Bajguz A (2011) Brassinosteroids—occurrence and chemical structures in plants. In: Hayat S, Ahmad A (eds) Brassinosteroids: a class of plant hormone. Springer, Dordrecht, pp 1–27
2. Tarkowská D, Novák O, Floková K et al (2014) Quo vadis plant hormone analysis? Planta 240:55–76
3. Swaczynová J, Novák O, Hauserová E et al (2007) New techniques for the estimation of naturally occurring brassinosteroids. J Plant Growth Regul 26:1–14
4. Takatsuto S, Ying B, Morisaki M et al (1982) Microanalysis of brassinolide and its analogues by gas chromatography and gas chromatography-mass spectrometry. J Chromatogr 239:233–241
5. Yokota T, Sato T, Takeuchi Y et al (2001) Roots and shoots of tomato produce 6-deoxo-28-norcathasterone, 6-deoxo-28-nortyphasterol and 6-deoxo-28-norcastasterone, possible precursors of 28-norcastasterone. Phytochemistry 58:233–238
6. Wu Q, Wu D, Shen Z et al (2013) Quantification of endogenous brassinosteroids in plant by on-line two-dimensional microscale solid phase extraction-on column derivatization coupled with high performance liquid chromatography-tandem mass spectrometry. J Chromatogr A 1297:56–63
7. Ding J, Mao L-J, Yuan B-F et al (2013) A selective pretreatment method for determination of endogenous active brassinosteroids in plant tissues: double layered solid phase extraction combined with boronate affinity polymer monolith microextraction. Plant Methods 9:13
8. Ding J, Wu J-H, Liu J-F et al (2014) Improved methodology for assaying brassinosteroids in plant tissues using magnetic hydrophilic material for both extraction and derivatization. Plant Methods 10:39
9. Tarkowská D, Novák O, Oklestková J et al (2016) The determination of 22 natural brassinosteroids in a minute sample of plant tissue

by UHPLC–ESI–MS/MS. Anal Bioanal Chem 408:6799–6812

10. Polko JK, Pierik R, van Zanten M et al (2013) Ethylene promotes hyponastic growth through interaction with *ROTUNDIFOLIA3/CYP90C1* in *Arabidopsis*. J Exp Bot 64:613–624

11. Singh AP, Fridman Y, Friedlander-Shani L et al (2014) Activity of the brassinosteroid transcription factors BRASSINAZOLE RESISTANT1 and BRASSINOSTEROID INSENSITIVE1-ETHYL METHANESULFONATE-SUPPRESSOR1/BRASSINAZOLE RESISTANT2 blocks developmental reprogramming in response to low phosphate availability. Plant Physiol 166:678–688

12. Stirk WA, Bálint P, Tarkowská D et al (2013) Hormone profiles in microalgae: Gibberellins and brassinosteroids. Plant Physiol Biochem 70:348–353

13. Stirk WA, Tarkowská D, Turečková V et al (2014) Abscisic acid, gibberellins and brassinosteroids in Kelpak®, a commercial seaweed extract made from *Ecklonia maxima*. J Appl Physiol 26:561–567

Chapter 2

Synthetic Protocol for AFCS: A Biologically Active Fluorescent Castasterone Analog Conjugated to an Alexa Fluor 647 Dye

Johan M. Winne, Niloufer G. Irani, Jos Van den Begin, and Annemieke Madder

Abstract

Synthetic derivatization of hormonally active brassinosteroids (BRs) can provide useful small molecule tools to probe BR signaling pathways, such as fluorescent analogs. However, most biologically active BRs are not suitable for direct chemical conjugation techniques because their derivatization typically requires extensive synthetic work and chemistry expertise. Here, we describe an operationally simple, two-step procedure to prepare and purify an Alexa Fluor 647-castasterone (AFCS) from commercially available materials. The reported strategy is also amenable to the introduction of various other amine-based labeling groups.

Key words Fluorescent labeling, Brassinosteroids, Castasterone, Steroid hormones, Oxime conjugation

1 Introduction

Fluorescent labeling of biologically relevant molecules is a valuable and classical strategy to probe biological processes [1]. Many reactive fluorophores have been developed and commercialized, such as the very popular fluorescein isothiocyanate, a reactive fluorescent dye that can covalently trap a range of chemical functions. Several commercial kits are now also available to perform standardized labeling of various substrate types, including some tailored for users with minimal experience in synthetic chemistry. However, for typical steroid hormones, most of the standard labeling strategies are hampered by a lack of suitable chemical functions, such as amines. Furthermore, steroids often contain multiple less reactive but similar functions within the same molecule, such as hydroxyls, precluding site-selective derivatization with "general" labeling methods. Moreover, because most functional groups in natural

Eugenia Russinova and Ana I. Caño-Delgado (eds.), *Brassinosteroids: Methods and Protocols*, Methods in Molecular Biology, vol. 1564, DOI 10.1007/978-1-4939-6813-8_2,

steroids are often essential for their bioactivity, their derivatization can easily result in loss of the hormonal function.

The plant steroid hormone, brassinolide (BL) (Fig. 1) is a particularly problematic steroid substrate for standard labeling methods. No less than four different but chemically similar hydroxyl groups are present at C-2, C-3, C-22, and C-23 that cannot be readily discriminated by typical chemical reagents. Moreover, all of these hydroxyl functions are directly implicated in the biological activity of BRs [2] and, thus, their derivatization is expected to result in significant loss of hormonal function. Furthermore, because the reactive lactone function at C-6 is an integral part of the BL secosteroid backbone, its derivatization through acylation reactions (lactone opening) also drastically alters the overall topology of the rigid "flat" molecular scaffold and influences its ability to be recognized by steroid receptors. In this regard, the biologically active precursor of BL, castasterone (CS) (Fig. 1) is a more attractive labeling substrate, because the ketone function at C-6 can be selectively modified without a priori risking major disruption of the receptor affinity. Moreover, CS is commercially available in multi-mg amounts at reasonable cost or can be synthesized from stigmasterol with a classical multistep sequence [3–5]. Previously, C-6 derivatives of CS have been shown to be able to retain their biological function as BR hormones [6].

In this chapter, we provide a detailed method for a small-scale derivatization of CS (approximately 1 mg) that should also be applicable for a number of related 6-oxo steroids. The method can be performed by researchers or technicians with minimal experience in synthetic chemistry. The protocol is a two-step derivatization strategy that goes through an oxime-conjugation step to introduce a carboxylic acid moiety into the CS framework that generates CS *O*-carboxymethyloxime (CS-CMO) (Fig. 2), followed by an amide coupling to an amine derivative of an Alexa Fluor (AF) 647 dye (AF 647 cadaverine). This strategy has also been employed to incorporate alternative tagging groups, such as

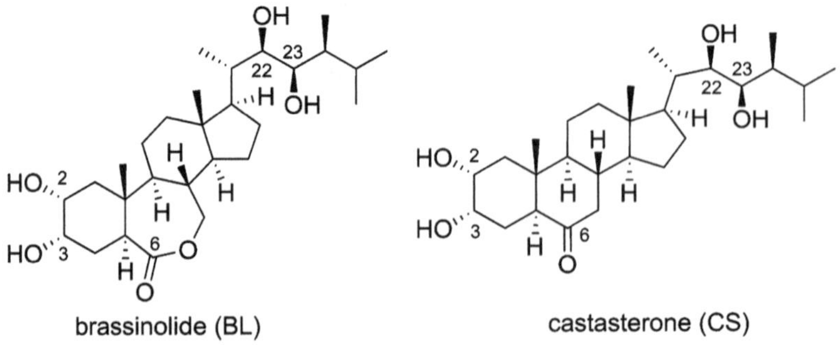

Fig. 1 Chemical structures of the hormonally active brassinosteroids brassinolide (BL) and castasterone (CS)

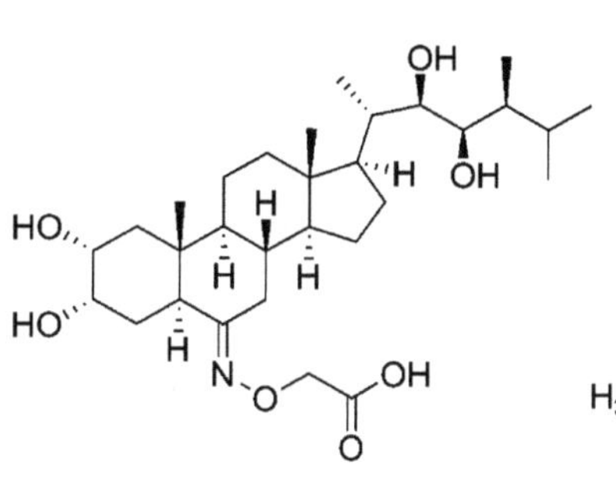

castasterone carboxymethyloxime
(CS-CMO)

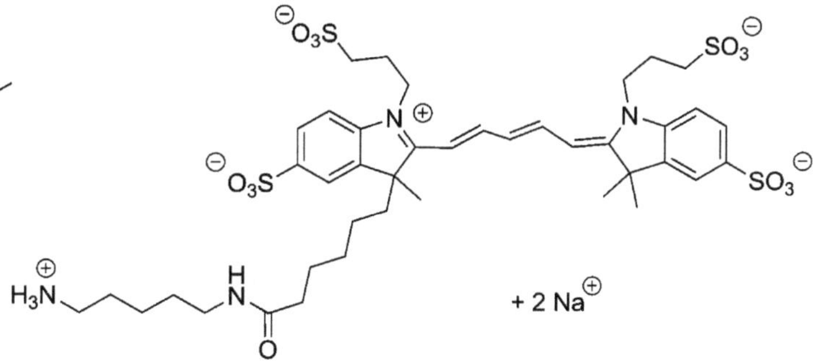

Alexa Fluor 647 cadavarine disodium salt

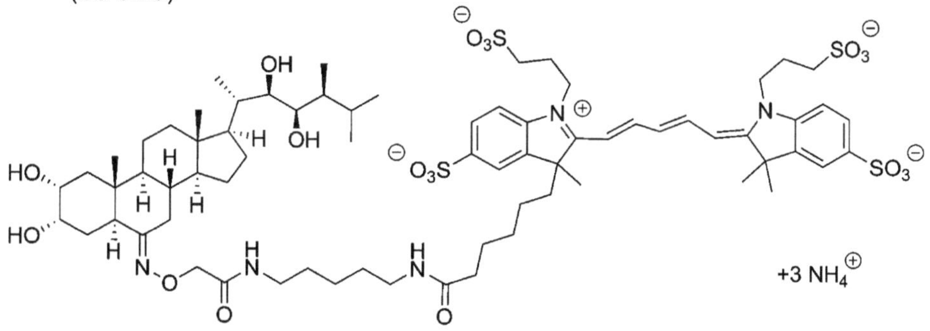

Alexa Fluor 647-castasterone conjugate (AFCS)

Fig. 2 Chemical structures of the C6-oxime derivative of castasterone, castasterone carboxymethyloxime (CS-CMO); an amine-derivative of Alexa Fluor 647 fluorescent dye, Alexa Fluor 647 cadaverine (AF 647 cadaverine); and of the derived CS–AF 647 amide-conjugate (AFCS)

a dansylcadaverine [6] or a BODIPY group [7]. Although a wealth of coupling reagents is available to affect an amide-type coupling between an amine and a carboxylic acid, the 4-(4,6-dimethoxy-1,3,5-triazin-2-yl)-4-methylmorpholinium chloride (DMT-MM) reagent used here is especially suitable for the envisaged small-scale derivatization, because of its high tolerance for water and operational ease [8].

2 Materials

2.1 Chemicals and Reagents

1. 5–10 mg powder of castasterone (CS) (CAS, 80736-41-0). CS can be prepared from stigmasterol via a number of well-established synthetic routes [3–5], but is also available from different chemical suppliers.
2. 1 mg powder of Alexa Fluor 647 cadaverine, disodium salt (AF 647 cadaverine; molecular mass approximately 1000 g/mol or 1 μmol). It can be purchased in 1-mg batches in individual screw capped plastic tubes and stored at −20 °C.

3. *O*-(carboxymethyl)hydroxylamine hemihydrochloride (CMHA) (CAS: 2921-14-4).
4. 4-(4,6-dimethoxy-1,3,5-triazin-2-yl)-4-methylmorpholinium chloride (DMT-MM) (CAS: 3945-69-5) This commercially available reagent should not be stored for very long times, because it loses activity over time by decomposition. The quality of the reagent is also known to vary. Fortunately, this problem can usually be solved by simply adding more reagent than required.
5. High-purity *N,N*-diisopropylethylamine (DIPEA) (CAS: 7087-68-5) (redistilled grade, >99 %) that can be used without further purification.
6. Solvents: methanol, glacial acetic acid, pyridine, acetonitrile, and chloroform, all at anhydrous high-performance liquid chromatography (HPLC) quality and used without further purification.

2.2 Chromatography Material

1. Standard silica gel for chromatography (60 Å).
2. C18-reversed phase silica gel for column chromatography (fully end-capped).
3. Standard glass-backed thin layer chromatography (TLC) plates with a conventional binder and fluorescent indicator.
4. Washed and dried sand for laboratory use.

2.3 Chromatography Solutions and Solvent Mixtures

1. LC solvent mixture: A mixture of chloroform (8.5 mL), methanol (1.5 mL), and glacial acetic acid (100 μL) prepared in a small volumetric cylinder and then transferred to a standard elution tank for TLC, i.e., a wide, flat-bottomed beaker with a cap to avoid evaporation.
2. TLC reference solution of CS: A CS solution in methanol (approximately 1 mg/mL) must be prepared in a small sealable vial or tube as a reference sample for TLC analyses (approximately 20 μL is enough).
3. A phosphomolybdic acid (PMA) TLC staining solution (10 % (v/v) in ethanol): 10 % (w/w) mixture of PMA hydrate (CAS: 51429-74-4) and ethanol (denatured technical grade).
4. A solvent mixture for chromatography elution: A mixture of 50 mL of chloroform, 2.5 mL of anhydrous methanol, and 250 μL of glacial acetic acid. This eluent is prepared, homogenized, and stored in a volumetric cylinder, sealed with a plug of laboratory tissue paper to avoid evaporation.
5. TLC solvent mixture: A mixture of chloroform (6 mL), methanol (4 mL), and glacial acetic acid (100 μL), prepared in a small volumetric cylinder and transferred to a standard elution

tank for TLC, i.e., a wide flat-bottomed beaker with a cap to avoid evaporation.

6. A solvent mixture for reversed-phase chromatography elution: A mixture of 33 mL of acetonitrile and 100 mL of a 5 mM aqueous solution of ammonium acetate (NH_4OAc), prepared by dissolving 38.5 mg of NH_4OAc in 100 mL of milli-Q deionized water.

2.4 Equipment and Other Materials

1. Glass Pasteur pipettes (approximately 2 mL internal volume) to transfer organic liquids, such as chloroform.
2. Micropipettes (ranges 1–20, 20–100, 100–500, and 500–2500 μL), suitable for organic solvents, with disposable tips to transfer and measure mixtures and solutions based on water, methanol, and/or acetonitrile.
3. High vacuum pump (down to 0.1 mm Hg).
4. Nitrogen line (purified nitrogen gas, 1–5 bar pressure regulator).
5. A rotary evaporator.
6. Cotton wool and paper towels.
7. Screw-capped glass reactor vials (1–10 mL volume) and Teflon-coated stirring bars that fit the internal diameter of the vials.
8. Glass fraction collection tubes (1–10 mL)
9. Microcentrifuge tubes.
10. A benchtop centrifuge.
11. A centrifugal concentrator.
12. An analytical balance, accurate to 0.1 mg.

3 Methods

Although the protocol has been written for researchers with minimal experience in organic synthesis, it is recommended to carry out the work in a well-equipped synthetic chemistry laboratory, with access to a fume hood, organic solvent handling and storage facilities, and analytical instruments, with minimal supervision of a trained organic chemist or laboratory technician.

3.1 Synthesis of CS-CMO

1. Weigh out CS (3.8 mg or 8.18 μmol) and CMHA (4.1 mg or 37.5 μmol) as powders (*see* **Note 1**) and place them together with a Teflon-coated stirring bar in a thick-walled screw-capped glass vial (volume 2 mL, approximately 0.5 cm internal diameter) (*see* **Note 2**). Dissolve the powders in approximately 0.25–0.30 mL of dry pyridine. Handle pyridine only in a well-ventilated fume hood to avoid exposure. Make sure no powder remains

on the sides of the reaction vessel. Immediately after addition of the dry pyridine, close the reaction vessel with a screw cap and stir the resulting mixture overnight at room temperature.

2. Sample the obtained reaction mixture (after the overnight stirring) by removing approximately 2 μL of the reaction mixture (with a micropipette or capillary) from the vessel. Transfer this liquid aliquot to a separate vial or tube. Inside a working fume hood, dry it under a small stream of nitrogen gas to remove most of the pyridine.
3. Redissolve the dried residue in approximately 25 μL of the LC solvent mixture. After homogenization of the resulting dilute solution, transfer a few μL to the baseline of a glass-backed TLC plate (with a capillary or micropipette tip), in a lane next to the applied similar amount of the TLC reference CS solution. Place the TLC plate upright in the elution tank containing the remainder of the LC solvent mixture, and allow it to absorb the eluent from its bottom (*see* **Note 3**). When the solvent front reaches the appropriate level, dry the TLC plate and visualize compounds under a UV lamp, by treatment with a PMA staining solution, and by heating (*see* **Note 4**). The starting CS material should appear as a spot with a ratio-to-front (Rf) value of approximately 0.53 (Fig. 3). The corresponding oxime CS-CMO appears as a lower running spot with an Rf value of approximately 0.23. Some minor spots may appear at Rf values above 0.53, including the remaining pyridine.
4. When the conversion of the starting CS material to the CS-CMO reaction product is incomplete, i.e., when it is still observed by the TLC analysis (Subheading 3.1, **step 3**) (Fig. 3), add additional CMHA (approximately 2 mg). Seal the reaction

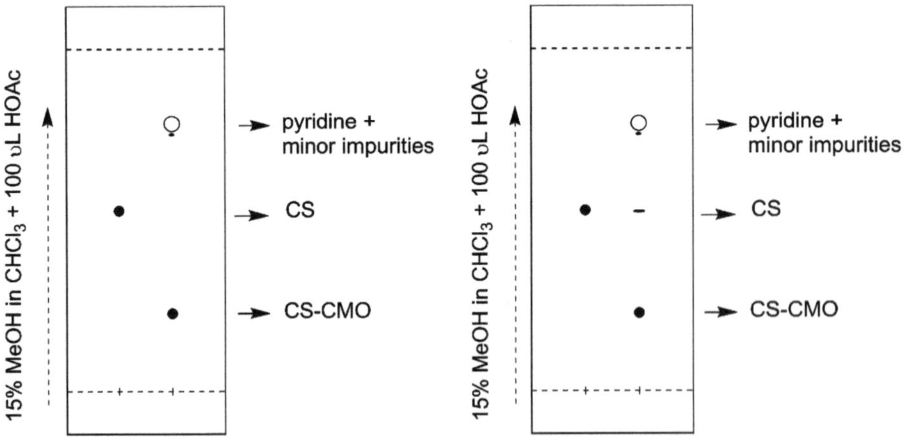

Fig. 3 Representation of thin-layer chromatograms (NP silica) for a complete oxime-forming reaction between CS and carboxymethylhydroxylamine (*left*), and that of an incomplete reaction between CS and carboxymethylhydroxylamine (*right*). All spots are non-UV fluorescent, except those at a very high Rf (pyridine)

vessel with the screw cap and then heat it to 60 °C for 90 min, with a complete consumption of all starting material as a result. Verify by TLC as described (Subheading 3.1, **steps 2** and **3**).

5. When a complete conversion to CS-CMO is finally confirmed by TLC analysis (Fig. 3), concentrate the reaction mixture under a small stream of nitrogen gas (*see* **Note 5**) and dry the residue on a high-vacuum pump.
6. Plug a glass Pasteur pipette (internal volume approximately 2 mL; *see* **Note 6**) with a small piece of cotton and then fill it with 500 mg of dry normal-phase silica gel (approximately 1 mL or half-height). Forcibly, wash the column filled with dry silica gel with 10–20 mL of the solvent mixture for chromatography elution with 1-mL portions. Use a nitrogen line or manual pump (pipette head) to apply pressure to the top of the column in order to remove any air from the silica during this rinsing. Be careful that no additional air is introduced into the silica column by maintaining the solvent line above the top of the silica column at all times (i.e., not pushing air through the silica). Discard the solvent that elutes during this rinsing and packing of the column. When the silica gel is visibly packed, i.e., semi-translucent without small air pockets or "white spots," allow the solvent line to drop to the top of the silica gel column (without pressure).
7. Dissolve the dried residue (obtained in **step 5**) in a pointed microcentrifuge tube, or reaction vessel that had been dried in a high vacuum, in approximately 50 μL of the solvent mixture for chromatography elution, and transfer the resulting solution immediately with a pipette to the top of the freshly packed silica gel column (prepared in **step 6**). From now on, collect the eluate of the column in a rack of small glass tubes (2–5 mL volume). Do not use plastic fraction collection tubes, because most are not compatible with the eluent. Rinse the sample tube containing the dried reaction mixture with approximately 50 μL of solvent, and transfer this solution to the top of the column. Finally, rinse the sides of the Pasteur pipette column with approximately100 μL of eluent. Then add a few mm of sand to the top of the silica gel column to avoid silica gel "whirling," before eluting the column by adding larger amounts of solvent (approximately1-mL portions). Continuously collect fractions of approximately 200–250 μL without letting the solvent line drop below the level of the sand. Elute the column with three times 1 mL of the initial solvent mixture, resulting in the collection of 10–15 fractions. Then, elute twice with 1 mL of the same solvent mixture after enriching it with 2 % (v/v) of methanol (*see* **Note 7**). Next, elute the column twice with 1 mL of the used solvent mixture, further enriched with 3 % (v/v) of methanol. At this point (around fraction 25), the oxime product

should start eluting from the column, as can be confirmed later by TLC analysis. Continue to elute four times with 1 mL of the same solvent enriched with another 5 % (v/v) of methanol.

8. Test the fractions collected in **step** 7 for the presence of CS-CMO by TLC analysis (Fig. 4). Pool the fractions that are positive for the desired compound, and also do not show any other products, because some minor mixed fractions may occur in the early fractions. Concentrate the combined CS-CMO-containing fractions with a rotary evaporator, in a vial or tube that has been carefully weighed beforehand. Redissolve the residue in a small amount of toluene (500 μL) and concentrate it again to dryness. Then, redissolve the residue in acetonitrile (500 μL) and evaporate to dryness three times (*see* **Note 8**). Finally, dry the residue under a high vacuum and weigh the vial to determine the amount of obtained product (approximately 5 mg; the maximum theoretical yield is 4.4 mg, i.e., near quantitative yield). The compound can be used without further purification in the next step. Standard analytical data (nuclear magnetic resonance or mass spectroscopy) can be checked and compared to literature data [6] to verify identity and purity.

3.2 Amide Coupling between CS-CMO and Alexa Fluor Cadaverine

1. Remove a 1-mL sealed plastic tube containing 1 mg AF 647 cadaverine disodium salt from the cold storage (−20 °C) and briefly allow it to warm to room temperature before use.
2. Dissolve the dried final compound CS-CMO (approximately 4.5–5.0 mg) (*see* Subheading 3.1, **step 8**) in 440 μL of anhydrous methanol (100 μL/mg). This solution will be used as CS-CMO stock solution (*see* **Note 9**).

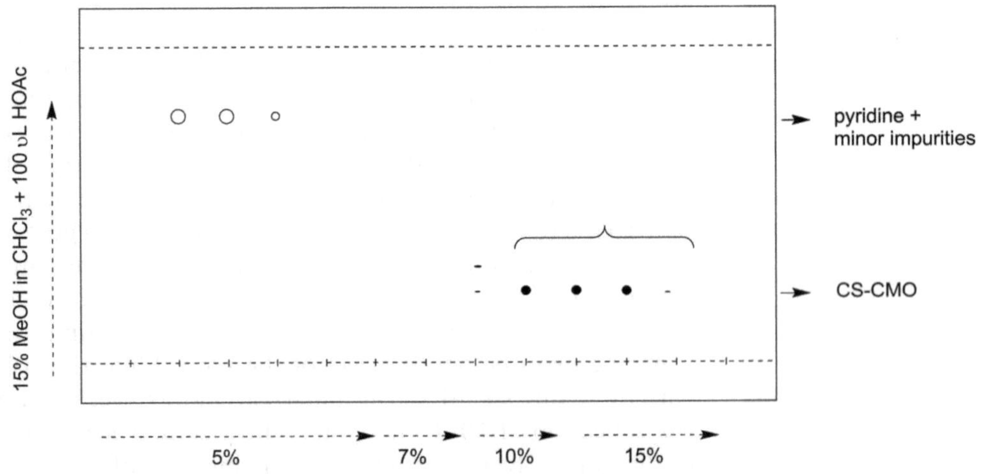

Fig. 4 Typical thin-layer chromatogram (NP silica) for fractions obtained from flash chromatography of the oxime-forming reaction between CS and carboxymethylhydroxylamine (CHMA), resulting in the CS-CMO compound

3. In a separate sealable vial or tube, prepare a homogenous stock solution of DIPEA in methanol by mixing 20 μL of DIPEA with 1980 mL of anhydrous methanol. This stock solution should be prepared directly before use to avoid or minimize contamination and evaporation and the excess should be discarded. Also weigh out two portions of approximately 1.3 mg of DMT-MM reagent (powder) in small sealable vials.
4. Open the tube containing 1 mg of AF 647 cadaverine disodium salt and fill it with 150 μL of the CS-CMO methanolic solution obtained in **step 2** (approximately 2.0–2.7 μmol). After homogenation by taking up and expelling the solution a couple of times in the tube with a micropipette, transfer the resulting bright blue solution to a similar reaction vessel as before (screw-capped 2-mL glass vial of approximately 0.5 cm diameter, *see* **Note 2**), equipped with a magnetic stirring bar. Rinse the AF 647 container tube with 20 μL of anhydrous methanol, and transfer this washing solution to the reaction vessel.
5. Add 60 μL of the freshly prepared DIPEA stock solution (approximately 2.7 μmol, *see* **step 3**) (*see* **Note 10**) to the reaction vessel already containing CS-CMO and AF 647 (*see* **step 4**). Next, add one portion of the DMT-MM reagent (1.3 mg, approximately 4 μmol) to the reaction vessel. Make sure all powder is dissolved in the reaction mixture, if needed rinse the sides with 20 μL methanol, before closing the vessel and stirring it magnetically with a Teflon-coated stir bar, for 10 min. At this time, another portion of DMT-MM reagent is added (1.3 mg) and the resulting mixture is stirred for another 20 min at ambient temperature.
6. Remove 1–2 μL of the reaction mixture obtained in **step 5** and transfer it directly to the bottom of a normal TLC plate. On the same plate, place a reference sample of both CS-CMO and AF 647 cadaverine in methanol, prepared by adding approximately 20 μL to the rinsed AF container tube of **step 4**. After running the TLC with the appropriate TLC solvent mixture, the conversion of the AF 647 cadaverine can be observed directly on the TLC plate, without visualization, by the disappearance of a bright blue spot at an Rf value of 0.11. A single bright blue new spot should appear at a significantly higher Rf value of 0.23 (Fig. 5, left) (*see* **Note 3**). After development with the PMA solution (*see* **Note 4**), some remaining CS-CMO compound, added in excess, can be observed near the top of the TLC plate (Rf 0.76). Upon observation of a complete reaction (*see* **Note 11**), concentrate the reaction mixture under a slow stream of nitrogen gas (*see* **Note 5**), and dry the residue under a high vacuum.
7. Purify the residue obtained in **step 6** by manual reversed-phase chromatography (*see* **Note 12**) with the solvent mixture for

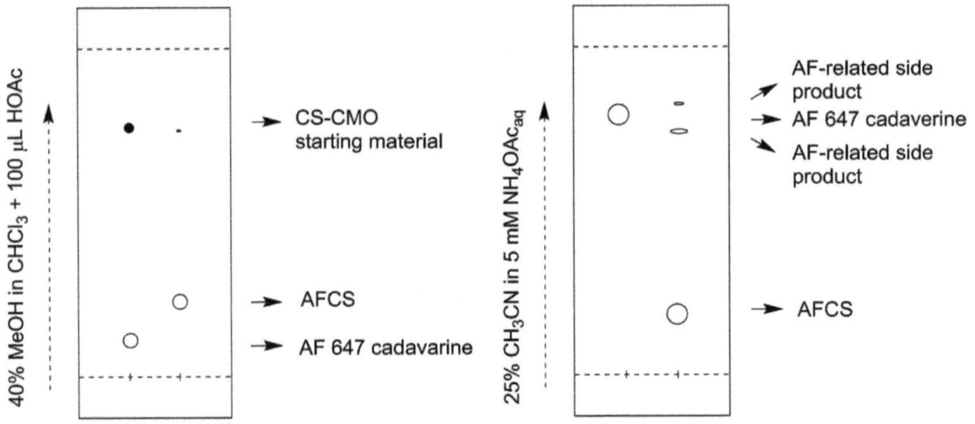

Fig. 5 Representation of a thin-layer chromatogram (*left*, normal-phase silica; *right*, reversed-phase silica) for a complete amide coupling reaction between CS-CMO and AF 647 cadaverine. All AF-related spots are colored bright blue. Upon overloading the TLC with compounds, the AFCS compound may show considerable tailing. Incomplete reactions typically show remaining AF 647 cadaverine and a complex "tail" of AF-related spots that clears up as the reaction proceeds

reversed-phase chromatography elution. Using this eluent in TLC analysis, the AFCS compound should have an Rf of approximately 0.2 on a reversed-phase silica gel TLC plate, whereas the impurities all have an Rf above 0.8 (Fig. 5, right) (*see* **Note 3**). Plug a standard laboratory-ware column (internal diameter 0.5–1 cm and a least 20 mm high) with a piece of cotton wool and fill it with approximately 2 g of slurried C18-reversed-phase silica with the same aqueous eluent. Rinse the column by forcibly washing it with the same eluent (*see* **Note 13**) until a good packing is achieved.

8. Dissolve the residue of **step 6** in 60 µL of aqueous eluent, and transfer the resulting bright blue homogenous solution to the top of the silica gel column. Rinse the vial with 30 µL of the aqueous eluent, and transfer this washing solution also to the top of the column. Rinse the sides of the column with 100 µL of eluent, and then further elute the column, while collecting fractions of approximately 200 µL, using either glass collection tubes or plastic microcentrifuge tubes. The progress of the chromatographic separation can be observed easily and directly by the appearance of two dark blue bands, one of which moves very rapidly down the column, virtually with the solvent front. This front running band should already elute in the first few fractions. As the second blue band can be observed to move slowly down the column, start collecting smaller and smaller fractions just before the blue band approaches the bottom of the column and starts to elute at a high concentration. Next, collect 100-µL fractions until the eluate is (almost) colorless. Analyze the lightly colored fractions before and after the dark

blue fractions by TLC to check purity. Pool all clean and AFCS-positive fractions, by using minimal amounts of eluent to rinse the fraction tubes.

9. Divide the combined clean AFCS-positive fractions from **step 8**, and distribute them equally over five microcentrifuge tubes (approximately 350 μL per vial). Next, centrifuge the tubes (*g* value?), freeze the spun down samples in liquid nitrogen, and then concentrate them under high vacuum with a centrifugal concentrator. Weigh the combined tubes before and after this operation to get an accurate determination of the isolated amount of AFCS: approximately 1.4 mg or 0.285 μg per tube (*see* **Note 14**). The identity of the AFCS compound can be confirmed by matrix-assisted laser desorption ionization-time of flight mass spectrometry.

4 Notes

1. It is important to get an accurate weight for the CS, because it can also be used as the basis for the calculations for reagents in the second step. For the added CMHA, a large excess is used, so its exact weight is less important. A 50 % relative deviation of the stated amount is acceptable and will give the same result.
2. The choice of reaction vessel is quite critical because of the small scale of the reaction. The reaction vessel should be air tight to avoid evaporation of solvent and exposure to moisture/oxygen. Classic round-bottomed flasks with a rubber septum are not suitable, because of the large surface area of the reaction mixture in such a vessel. In the absence of a straight-walled glass pressure tube, a plastic sealed microcentrifuge tube can also be acceptable.
3. The outcome of the TLC analysis can be influenced by the presence of the solvent remaining from the compound application (methanol or pyridine). Therefore, the spots on the TLC plate could be dried for 1 min under a stream of nitrogen gas, and then left to dry for 10 min before transfer to the elution tank. Also make sure the solvent line in the elution tank is well below the line on the TLC plate where compounds are applied.
4. Other common laboratory TLC staining reagents will also give the desired result, such as anisaldehyde- or permanganate-based stains.
5. Alternatively, the reaction mixture can be concentrated with a standard rotary evaporator. On this small scale, however, simple insertion of a Pasteur pipette connected to a source of nitrogen gas halfway in the reaction vessel will effectively and quickly remove all volatiles in a matter of minutes. The evapo-

ration can be helped by gently heating the vessel with your hands. The stream of nitrogen should be as fast as possible without risking splashing of the solvent.

6. A standard laboratory-ware column can also be used, but because of the usually larger internal diameter, more silica gel is required, resulting in a higher loss of compound through irreversible absorption. A Pasteur pipette is most convenient and its small dead volume at the exit can be further reduced by breaking off the tip just below the cotton plug. A disadvantage of the use of a pipette is that the chromatographic separation cannot be halted at any time, but because of the small scale, should not be an important problem. However, all handlings have to be done swiftly without delay to avoid "drying out" of the column.
7. Keeping the solvent mixture for column elution in a 50-mL measuring cylinder allows rapid gradient adjustment by simple addition of the required amount of additional methanol and homogenization of the resulting mixture.
8. This coevaporation with toluene and acetonitrile is absolutely required to remove all residual acetic acid from the compound, because it will also be reactive to the AF 647 cadaverine and will thus severely limit the efficiency of the conjugation reaction.
9. As the CS-CMO compound is not a powder, a stock solution in methanol is the most convenient way to control the quantity of the compound at this scale. The method also does not require accurate weighing of very small amounts, which can be tedious. The unused stock solution can be concentrated and dried again and stored for future use.
10. The addition of DIPEA is usually not required for DMT-MM amide couplings, but is necessary in this case because the commercial AF 647 cadaverine compound is in fact an internal ammonium sulfonate salt that needs to be deprotonated before it can be acylated. To this end, addition of this non-nucleophilic amine in an equimolar ratio to the CS-CMO compound is the most convenient.
11. When some AF 647 starting material can still be seen on the TLC plate, it is most probably due to partial decomposition of the labile DMT-MM reagent, because the amide formation reaction should be fast. In this case, the reaction can be driven to completion by addition of 1–2 mg more DMT-MM reagent and by stirring the resulting mixture for another 30 min. Do not add additional DIPEA.
12. A semi-preparative automated HPLC purification can also be carried out, but the manual chromatography is very easy and convenient to perform because of the bright blue color of both

the desired compound and side products. Manual normal-phase chromatography can also be used to purify AFCS, but the acidic medium can result in partial methanolysis of the AFCS compound during chromatography, as can already be observed as a light blue "tailing" of the AFCS spot, making this technique less suited. Moreover, the separation is very straightforward on reversed-phase silica, because all AF-based contaminants have a significantly shorter retention time and the excess CS elutes much later.

13. Some inconvenience may arise from the fact that the density of (dry) reverse-phase silica gels is lower than that of the solvent mixture. Slurrying the silica gel before loading the column should help, and rinsing with a lot of solvent for a long time under high pressure should give a tightly packed column. This problem is also the reason why performing the manual flash chromatography on a Pasteur pipette (*see* **Note 6**), is not a practical option in this case. A wider internal diameter is needed to allow the aqueous slurry to accumulate at the bottom of the column.
14. A more accurate quantification of AFCS can be done on solutions prepared from these tubes by standard colorimetric methods.

Acknowledgments

NGI is indebted to the Odysseus program of the Research Foundation-Flanders for a postdoctoral fellowship.

References

1. Gonçalves MST (2009) Fluorescent labeling of biomolecules with organic probes. Chem Rev 109:190–212
2. Sun Y, Han Z, Tang J (2013) Structure reveals that BAK1 as a co-receptor recognizes the BRI1-bound brassinolide. Cell Res 23:1326–1329
3. Back TG, Blazecka PG, Krishna MV (1993) A new synthesis of castasterone and brassinolide from stigmasterol. A concise and stereoselective elaboration of the side chain from a C-22 aldehyde. Can J Chem 71:156–163
4. Honda T, Keino K, Tsubuki M (1990) A concise stereoselective synthesis of castasterone. J Chem Soc Chem Commun 8:650–652
5. Kametani T, Katoh T, Fujio J et al (1988) An improved synthesis of plant growth regulating steroid brassinolide and its congeners. J Org Chem 53:1982–1991
6. Irani NG, Di Rubbo S, Mylle E et al (2012) Fluorescent castasterone reveals BRI1 signaling from the plasma membrane. Nat Chem Biol 8:583–589
7. Malachowska-Ugarte M, Sperduto C, Ermolovich YV et al (2015) Brassinosteroid-BODIPY conjugates: design, synthesis, and properties. Steroids 102:53–59
8. Kunishima M, Kawachi C, Morita J et al (1999) 4-(4,6-dimethoxy-1,3,5-triazin-2-yl)-4-methyl-morpholinium chloride: an efficient condensing agent leading to the formation of amides and esters. Tetrahedron 55:13159–13170

Chapter 3

Physiological Analysis of Brassinosteroid Responses and Sensitivity in Rice

Hongning Tong and Chengcai Chu

Abstract

Identification of related mutants contributed a lot to our understanding of the molecular mechanisms underlying brassinosteroid (BR) signaling. Establishment of methods to test for the hormone sensitivity is essential for the characterization of the mutants associated with BR responses. Here, we describe detailed procedures for the physiological analysis of BR responses that are regularly applied to test the BR sensitivity of mutants by evaluating the growth of different tissues of rice (*Oryza sativa*) seedlings.

Key words Brassinosteroid sensitivity, Brassinosteroid response, Lamina inclination, Coleoptile elongation, Rice

1 Introduction

As a class of growth-promoting phytohormones, brassinosteroids (BRs) play important and diverse roles in regulating plant growth and development in response to various environmental conditions. Since its discovery approximately 40 years ago, biological processes of the hormone biosynthesis and signaling have been intensely studied, especially in the model plant *Arabidopsis thaliana*. As the hormone has been demonstrated to have a great potential to improve plant performance and productivity, further studies in crop plants, such as rice (*Oryza sativa* L.), have particular significance for application in the field. However, the monocotyledonous plant rice has a very different plant architecture as well as growth conditions from the dicotyledonous plant *Arabidopsis*. Thus, establishment of feasible methods for the physiological analysis of the BR responses and sensitivity in rice will benefit further BR studies. Here, we describe our routine protocols for the physiological analysis of BR responses and sensitivity in rice seedlings. Compared to the previous similar methods [1–3], the modified procedures described here are more feasible and convenient to perform under laboratory conditions. The analyses are normally applied to

Eugenia Russinova and Ana I. Caño-Delgado (eds.), *Brassinosteroids: Methods and Protocols*, Methods in Molecular Biology, vol. 1564, DOI 10.1007/978-1-4939-6813-8_3, © Springer Science+Business Media LLC 2017

compare a mutant seedling growth situation to that of its wild-type plant with or without the hormone application. The plants are cultured with basal medium under regular light and temperature conditions. We include four tissues for the tests, but it should be noted that, whereas BR promotes lamina inclination and coleoptile elongation, the leaf and the root could be repressed by exogenous brassinolide (BL) [4]. As different tissues of rice have different responses to BR treatments, the results obtained from these assays should be appropriately explained when making a conclusion.

2 Materials

1. 20 mM (10 μg/μL) brassinolide (BL) stock solution: add 1 mL of dimethyl sulfoxide (DMSO) to the bottle of the original package containing 10 mg BL. Dissolve and split into aliquots in 100-μL tubes. Store at −20 °C.
2. ½ Murashige and Skoog (MS) nutrient solution: dissolve 2.2 g MS basal medium powder in 1 L of deionized water. Prepare before use.
3. ½ MS solid culture medium: add 3.2 g phytagel to 1 L of ½ MS nutrient solution, autoclaved for preservation.
4. Sterilization solution: dilute NaClO (containing 10 % (v/v) effective chlorine) fifty-fifty with deionized water.
5. Disposable 96-well polymerase chain reaction (PCR) plate.
6. Rectangular plastic box: the top size should match or be larger than the 96-well PCR plate.
7. Cylinder glass tissue culture flask (5.5 cm diameter, 18 cm height).
8. Growth chamber, with controllable light and temperature.
9. Clean laminar-flow bench.
10. Mini-type huller (optional).
11. Filter papers, petri dishes, conical flasks, drying oven, and other regular laboratory equipment.

3 Methods

3.1 BR Sensitivity Assay by Evaluating the Lamina Inclination in Response to BL

Lamina joint is a specialized tissue consisting of a bunch of cells that connect leaf sheath and leaf blade. Lamina inclination was thought to be a BR-specific response because it was tightly regulated by BR. BR simultaneously inhibits the abaxial cell division and promotes the adaxial cell expansion of the lamina joint to enhance lamina inclination [5]. Thus, the lamina inclination assay is the most popular method used to test the BR sensitivity in rice.

The description below was applied to compare the BR sensitivity between two plant lines (such as mutant versus wild type).

1. Surface sterilize sufficient rice seeds (*see* **Note 1**) by using 25 mL of the sterilization solution in a 50-mL conical flask for 30 min with shaking (*see* **Note 2**).
2. Wash the seeds for twice with water and then immerse the seeds in water for 6 h at room temperature (*see* **Note 3**).
3. Change to fresh water and put the flask in a growth chamber at 37 °C for two additional days until the seeds germinate with a radicle length of approximately 1 mm (*see* **Note 4**). Rinse the seeds once during the incubation.
4. Cut off the bottom of all the wells (approximately 0.3 cm length) of the 96-well PCR plate with sharp scissors. The bottom holes should be suitable to hold a seed and allow the root to grow downward (*see* **Note 5**).
5. Put the plate on the rectangular box (*see* **Note 6**). This setup will be used to culture the rice seedlings, with the plate wells to sow the germinated seeds and the box to fill with the nutrient solution.
6. Select the germinated seeds with an identical radicle length. Place them into the wells one by one with the embryo upward (*see* **Note 7**), and fill the box with the ½ MS nutrient solution, of which the liquid level should reach the seed radicle.
7. Put the setup (Fig. 1) in a growth chamber with settings a 10-hour light–14-hour dark regime and 30 °C (*see* **Note 8**). During growth, replenish with some water when the liquid level gets too low.
8. After approximately 3 days of growth, the second leaf will grow out with the vertical and slightly expanded leaf blade (Fig. 1). This is the time for the BL application (*see* **Note 9**).
9. Dilute the BL storage solution with ethanol to a concentration gradient of 1000, 100, 10, and 0 ng/μL with a volume of approximately 50 μL for each concentration.
10. Drop 1 μL of the solution to the adaxial side of the leaf blade tip (slightly downward; Fig. 1) using a pipette (*see* **Note 10**). For each concentration, 12 plants (one row) are used for the test.
11. After an additional 3-day growth, take out the plants and spread them on a sheet to take a digital photo. The leaf angles can be measured on a computer with the free software ImageJ (http://imagej.net/).
12. For data analysis, calculate the average values for each concentration and compare with those of the wild type (*see* **Note 11**).

Fig. 1 Comparison of the BR sensitivity of wild type (*WT*) and *dwarf and low-tillering* (*dlt*) mutant by using the lamina inclination test in response to BL. The setup for plant cultivation is shown. *Blue arrows* indicate the position for the BL application and *arrowheads* the lamina joint

3.2 BR sensitivity Assay by Evaluating the Growth Curve of Coleoptile, Leaf, and Root in Response to BL

Coleoptile is another tissue that is very sensitive to BR. Exogenous BR can effectively promote coleoptile elongation. However, in leaves and roots, BR can either inhibit or slightly promote or not affect their growth, depending on the plant backgrounds, hormone concentrations, application time, and other effects [4]. The method below is applied to compare the BR sensitivity between two plant lines (such as mutant versus wild type). Aseptic conditions should be employed as indicated.

1. Prepare for autoclaving (*see* **Note 12**): 16 tissue culture flasks (with cover or sealed with paper and rubber band), four pieces of filter papers (9 cm), two glass petri dishes (9 cm, with cover), deionized water (400 mL), and two conical flasks (50 mL, sealed with paper and rubber band). Autoclave, dry in a 60 °C oven, and keep for further use.
2. Dilute the BL to a concentration gradient in 100 μL DMSO by 10:1 gradient dilutions, resulting in: 2×10^{-2} M (the storage solution), 2×10^{-3}, 2×10^{-4}, 2×10^{-5}, 2×10^{-6}, 2×10^{-7}, 2×10^{-8}, and 0 M (without BL).
3. Prepare eight bottles of 150 mL ½ MS culture medium for each concentration of BL by melting the solid medium in a microwave oven (*see* **Note 13**). Cool down naturally to approximately 60 °C (experimentally the bottle can be held in bare hands for a while) or put in a 60 °C oven for a longer time.

4. In a clean laminar-flow bench, add 75 μL of BL prepared in the last step to the medium, one bottle with one concentration. The final concentrations of the BL are 10^{-5}, 10^{-6}, 10^{-7}, 10^{-8}, 10^{-9}, 10^{-10}, 10^{-11}, and 0 M.
5. Shake vigorously in one direction to avoid bubbles. Pour one bottle of the medium into two culture flasks with equal volume (75 mL) immediately. Wait for the medium to cool down and solidify before use.
6. Select carefully fully-filled rice grains of the mutant and wild type and dehusk with a mini-type huller or by hand. The dehusked seeds are selected again to collect identical ones with good appearance (*see* **Note 14**).
7. Sterilize at least 120 seeds (15 seeds for each BL concentration) for each plant (mutant or wild type) in the sterilization solution for 30 min in the sterilized conical flasks with shaking (*see* **Note 15**).
8. Discard the sterilization solution. In the bench, put the two layers of the filter papers in the petri dish and then spread the seeds onto the papers. Air dry the seeds in the bench (*see* **Note 16**).
9. Transfer the seeds onto the medium (15 seeds for each flask). Adjust the seed position in order to uniformly spread them on the medium surface with a long forceps (*see* **Note 17**).
10. Seal the flasks and put them in a growth chamber under regular growth conditions (10-hour light–14-hour dark, 30 °C; *see* **Note 8**). Grow for one week.
11. Take the plants out, and measure the lengths of the coleoptile, root, and second leaf sheath (*see* **Note 18**).
12. Draw the growth curve of the different tissues in response to various BL concentrations and compare with the wild type (*see* **Note 19**).

4 Notes

1. Approximately 50 identical germinated grains for each mutant or wild type are needed for one assay. As mutant seeds normally tend to have a decreased germination rate or delayed germination, enough seeds should be prepared to ensure that sufficient identical germinated seeds could be selected for further analysis. In this case, the solution volume could be increased accordingly.
2. Surface sterilization can effectively enhance the seed germination rate and also decrease other potential effects on seedling growth.

3. This imbibition under relatively low temperature will allow the seeds to absorb sufficient water at the first stage, which can increase the consistency of the germination time.
4. Depending on seed status and rice background, the germination time could differ from 1 to 3 days.
5. We suggest the use of the 96-well PCR plate to culture the rice seedlings in the lamina inclination assay, because it guarantees identical plant intervals, therefore decreasing variations. Any other device can be used to grow the plants.
6. We utilize the plastic preservation box (without the lid) or the disposable tip box (the bottom part), because both their size perfectly fits the 96-well PCR plate. If the box tends to be larger, twine it with several rubber bands and put the plate onto the bands to fix the plate.
7. Make sure identical seeds are used in the test because this is critical for the assay. Mutant seeds can be sown on one half of the plate (12×4 or 6×8) and the wild-type seeds on another half (Fig. 1).
8. As light could have a great effect on BR sensitivity, different settings of the light cycle will affect the result, but should not be critical for the sensitivity comparison.
9. Depending on the rice background, the growth time could differ. Keep observation on the leaf blade status, because a slightly expanded leaf blade is more favorable for application of the BL drop on the leaf blade. Keep in mind that the first rice leaf is an incomplete leaf without blade differentiation.
10. As ethanol is volatile, drop the solution quickly and keep the tube closed when not used.
11. In most cases, wild-type plants have a leaf angle of approximately 90° when 100 ng BL is applied. For example, see [6].
12. This material is prepared for two lines, one mutant and its wild type. Prepare more accordingly if you need to test more plants.
13. Can also be freshly prepared. More medium should be made if you want to test more seeds.
14. Uniformity of the seed germination is critical for the result. Thus, try to pick the best ones for the test.
15. In physiological analysis, we always encourage to test more samples to achieve more accurate results. Thus, additional seeds can be tested by preparing more medium in more flasks.
16. Direct drying of the seeds without washing can effectively decrease the possible contamination by microbes.
17. Do not push the seeds downward too much, because gel embedding of the seeds will greatly promote coleoptile growth.

Adjust seed posture only by a slight touch, because the medium is soft enough to sustain the seeds.

18. Alternatively, seedling height can be measured to evaluate the effect, which usually has a similar growth curve tendency as the leaf sheath. You may need to wash the root to remove the agar before measurement.
19. For example, see [4]. Note that BL inhibits leaf and root growth at high concentrations.

Acknowledgments

This work was supported by grants from National Natural Science Foundation of China (91435106, 91335203, and 31170715) and Youth Innovation Promotion Association, Chinese Academy of Sciences (2015076 to H.T.).

References

1. Wada K, Marumo S, Abe H et al (1984) A rice lamina inclination test—a micro-quantitative bioassay for brassinosteroids. Agric Biol Chem 48:719–726
2. Fujioka S, Noguchi T, Takatsuto S et al (1998) Activity of brassinosteroids in the dwarf rice lamina inclination bioassay. Phytochemistry 49:1841–1848
3. Hong Z, Ueguchi-Tanaka M, Umemura K et al (2003) A rice brassinosteroid-deficient mutant, *ebisu dwarf* (*d2*), is caused by a loss of function of a new member of cytochrome P450. Plant Cell 15:2900–2910
4. Tong H, Xiao Y, Liu D et al (2014) Brassinosteroid regulates cell elongation by modulating gibberellin metabolism in rice. Plant Cell 26:4376–4393
5. Sun S, Chen D, Li X et al (2015) Brassinosteroid signaling regulates leaf erectness in *Oryza sativa* via the control of a specific U-type cyclin and cell proliferation. Dev Cell 34:220–228
6. Tong H, Liu L, Jin Y et al (2012) DWARF AND LOW-TILLERING acts as a direct downstream target of a GSK3/SHAGGY-like kinase to mediate brassinosteroid responses in rice. Plant Cell 24:2562–2577

Chapter 4

Light Regulation of Brassinosteroid Signaling Components: Checking Regulation of Protein Stability in Darkness

Claudia Corvalán and Sunghwa Choe

Abstract

Environmental conditions can affect stability of proteins at transcriptional or posttranscriptional levels to modulate their functions. Here we describe a method to observe changes in protein stability under different light conditions. In brief, *Arabidopsis thaliana* seedlings were maintained under various light regimes from continuous light to total darkness or transitions from light to dark, whereafter total protein was extracted from plants. Proteins were measured and resolved on sodium dodecyl sulfate–polyacrylamide gels and transferred to polyvinylidene difluoride membranes. Blots were incubated with the corresponding antibodies for the visualization of protein bands. The protocol described has been successfully applied in wild-type, different transgenic, and mutant background plants to study how light alone or in combination with other factors influences protein stability.

Key words *Arabidopsis thaliana*, Dark treatments, Light treatments, Protein extraction, Immunoblot, Antibody

1 Introduction

Growth under light or in total darkness results in photomorphogenesis (deetiolation) or skotomorphogenesis (etiolation), respectively. Light regulates morphogenesis by interactions with various signaling pathways involving different phytohormones. Multiple evidence was found that brassinosteroids (BRs) play a role in light-regulated development, of which one of the first were the light-grown phenotypes displayed in the dark-grown BR-deficient mutants *deetiolated-2* (*det2*) and *constitutive photomorphogenesis and dwarfism* (*cpd*). Additionally BR, mainly through the key transcriptional factor BRASSINAZOLE RESISTANT1 (BZR1), exerts transcriptional regulation of important components of the light signaling pathway [1–4]. Still the molecular mechanisms and components regulating BR and light signal interactions are far from being elucidated, although a few studies about protein interactions answer some of these questions [5–7].

Eugenia Russinova and Ana I. Caño-Delgado (eds.), *Brassinosteroids: Methods and Protocols*, Methods in Molecular Biology, vol. 1564, DOI 10.1007/978-1-4939-6813-8_4, © Springer Science+Business Media LLC 2017

The methods used until now include mainly comparisons between two conditions: plants grown under continuous light or darkness [8–10]. Although examination of these conditions was informative, it does not necessarily reflect the changes that occur at the protein level when light conditions change. For instance, a previous work showed almost no differences in the BZR1 phosphorylation status, which determines the BZR1 activity, among dark- and light-grown seedlings [9]. However, we found that changes in light for short dark treatments of just few hours could modify the protein stability and phosphorylation status [5]. The transition from light to dark allows examination of changes at the protein level by immunoblot assays or at the mRNA level by microarray or quantitative real-time reverse-transcription-polymerase chain reaction (RT-PCR).

Total protein is extracted from plants for a wide range of applications, including protein separation, peptide mapping, identification, and characterization. Immunoblotting allows the detection and characterization of proteins from total extracts in several steps [11] that are presented in this book chapter. The first part corresponds to the preparation of protein extracts from plants that differs from protocols used in animals because of the presence of the cell wall that must be sheared to release the cell contents [12]. Many recipes of extraction buffers are available and choosing the appropriate buffer to release of the protein of interest is important and will depend on the ability to solubilize the protein from its localization (whole cell, cytoplasmic, membrane-bound, nuclear, etc.). The variables include salt concentration (that varies from 0 to 1 M), type and amount of detergent, presence of divalent cations, and pH (between 6 and 9). Most extraction buffers contain sodium dodecyl sulfate (SDS) or other ionic detergents to increase the protein solubility, but it is considered a harsh denaturing buffer. Hence, we prefer not to use it because it may affect the protein quantification.

For our extraction buffer, we use less denaturing nonionic detergent, such as glycerol, Triton X-100, and NP-40 [12, 13]. Once lysis occurs, proteolysis, dephosphorylation, and denaturation begins. Therefore, we prefer to maintain the samples in cold, because temperature affects the catalytic activity of proteases, and we add fresh protease and phosphatase inhibitors in this protocol. We include ethylenediaminetetraacetic acid (EDTA) along with Mg^{2+} ions ($MgCl_2$) as cofactors to inhibit metalloproteases and phenylmethylsulfonylfluoride (PMSF), because many plant proteases are of the serine type [12]. Finally, dithiothreitol (DTT), as a reducing agent to prevent oxidation of proteins, could lead to protein aggregation [14]. The addition of all these components to the buffer is useful for the stability of the proteins, which can also be stored at −20 °C or −80 °C for a few weeks. Following extraction, proteins are separated by polyacrylamide gel electrophoresis

(PAGE) and are electrophoretically transferred to a polyvinylidene difluoride (PVDF) membrane. Finally, the proteins on the membranes are subjected to antibody binding for detection [15].

Using this protocol, we were able to detect differences in the protein phosphorylation status and stability after relatively short time changes in the light conditions. The described light treatments have been also applied in combination with whole transcriptome analysis as well in gene expression studies by real time RT-PCR. Additionally, protein interaction analyses, such as immunoprecipitation, were also effectively accomplished through a method based on this protocol.

2 Materials

2.1 Seedling Treatment

1. Agar-solidified Murashige and Skoog (MS) medium: 2.2 g/L MS salts, 10 g/L sucrose, 7 g/L agar, pH 5.75 (adjust with KOH solution).
2. Seven-day-old *Arabidopsis thaliana* seedlings.
3. Liquid nitrogen (N_2).

2.2 Protein Extraction

1. Protein extraction buffer: 50 mM Tris–HCl (pH 7.5), 100 mM NaCl, 10 mM $MgCl_2$, 1 mM EDTA, 10 % (v/v) glycerol (*see* **Note 1**).
2. Additives to the extraction buffer: 1 mM DTT, 1 mM PMSF, 1× protease inhibitor cocktail (*see* **Note 2**).

2.3 Protein Measurement

1. Protein Assay Dye Reagent Concentrate (5×) for protein determination (Bradford solution).
2. 100× (10 mg/mL) bovine serum albumin (BSA).
3. Spectrophotometer.

2.4 Immunoblot Analysis

1. SDS-PAGE running gel: 10 % (w/v) acrylamide mix, 1.5 M Tris–HCl (pH 8.8), 10 % (w/v) SDS, and 10 % (w/v) ammonium persulfate (APS) and *N,N,N′,N′*-tetramethylethylenediamine (TEMED).
2. SDS-PAGE stacking gel: 4 % (w/v) acrylamide, 1 M Tris–HCl (pH 6.8), 10 % (w/v) SDS, and 10 % (w/v) APS and TEMED.
3. Tris–glycine SDS buffer (running buffer): 25 mM Tris, 192 mM glycine, 0.1 % (w/v) SDS.
4. 5× SDS sample buffer: 10 % SDS (w/v), 10 % (v/v) 2-mercaptoethanol or 0.5 M DTT, 50 % (v/v) glycerol, 0.25 % (w/v) bromophenol blue, 0.25 M Tris–HCl (pH 6.8).
5. PVDF membrane.

6. Tris–glycine buffer (transfer buffer): 25 mM Tris, 192 mM glycine, 20 % (v/v) methanol (*see* **Note 3**).
7. Phosphate buffered saline (PBS): 27 mM KCl, 1.37 M NaCl, 100 mM Na_2HPO_4 and 18 mM KH_2PO_4.
8. Blocking buffer: 5 % (w/v) skim milk in 1× PBS.
9. First antibody against the protein of interest and horseradish peroxidase (HRP)-conjugate secondary. The antibodies are prepared in 1 % (w/v) skim milk in 1× PBS and the amount of antibody required is usually indicated by the provider.
10. Chemiluminescent Western Blot Detection kit.
11. Any gel documentation system that supports chemiluminescent application, such a MicroChemi System or ChemiDoc™.

3 Methods

3.1 Seedling Treatments

1. To perform the light treatments, grow *Arabidopsis* seedlings. Surface sterilize seeds and plate them on MS medium. After stratification at 4 °C for 3 days, grow seedlings under long-day (16-h light–8-h dark) light conditions with a photon fluence rate of 80 μmol m^{-2} s^{-1} at 22 °C for 7 more days before treatments (*see* **Note 4**).
2. For each light regime to be tested, use one plate of seedlings. We defined four different conditions; cL (continuous light), LD (light-to-dark transition), LDL (light-to-dark and return to light), and cD (continuous dark) (*see* **Note 5**).
3. For the cL regime, maintain seedlings under the same light conditions they originally grew. We refer to this regime as the control.
4. For LD conditions, transfer plates with seedlings to total darkness for 6 h. Place the plates in the same growth room along with the control and other conditions and simply wrap them with three layers of aluminum foil (L → D 6 h).
5. For the LDL regime, transfer plates to darkness (LD) and shift back to light for additional 6 h (LD 6 h → L 6 h).
6. For the cD regime, keep plates from the beginning, right after stratification, under darkness. Check that seedlings present the typical skotomorphogenic development, which is etiolated, with long hypocotyls, and small cotyledons.
7. At the end of each treatment, collect tissue from whole seedling as quickly as possible to avoid stressful conditions and place it in liquid N_2. Tissues can be stored at −80 °C when necessary. Tissue of the seedlings under darkness at the sampling time must be collected in a dark room.

3.2 Protein Extraction

1. Grind the tissues (at least 200 mg of whole seedlings) using mortar and pestle under liquid N_2 and collect the powder in frozen tubes. Keep tubes in liquid N_2 until the end of the preparation of all the samples.
2. Remove the tubes from the liquid N_2 and add 80–100 μL of extraction buffer or an amount enough to resuspend all the powder from the ground tissue (*see* **Note 6**).
3. Keep the samples on ice for approximately 10 min.
4. Centrifuge at 17,000 × *g* for 15 min at 4 °C with a microcentrifuge.
5. Rescue the supernatant in a new tube and keep samples on ice (*see* **Note 7**).

3.3 Protein Measurement

To determine the protein concentration, the method uses BSA as a reference protein to compare the values of the samples to that of BSA, of which the concentration is known.

1. Prepare dilutions of BSA to obtain six standard points (i.e., 0, 2, 4, 6, 8, and 10 μg/μL).
2. Mix 1 mL of 1× Bradford solution with 1 μL of protein samples, including the six standard points, and let at room temperature for 5–10 min (*see* **Note 8**).
3. Measure the absorbance (color intensity) of all samples at 595 nm in a spectrophotometer.
4. Generate a linear regression for the set of standards, and calculate the protein concentrations of the samples using the equation for this line (*see* **Note 9**)

3.4 Immunoblot Analysis

1. Mix the desired amount of total protein with 5× SDS sample buffer (*See* **Note 10**).
2. Boil the sample for 7 min and cool down before loading (*see* **Note 11**).
3. Load the samples onto 10 % SDS-PAGE gel and run with 1× running buffer until dye reaches the bottom of the gel.
4. Transfer the proteins in the gel to the PDVF membrane using transfer buffer (*see* **Note 12**).
5. Once the transfer is finished, put the membranes for 1–2 h in blocking buffer with continuous agitation (50 rpm).
6. Discard the blocking buffer and replace it with the first antibody, which is the antibody against the protein of interest. Incubate the membrane for 2 h in the solution with the antibody under continuous agitation (50 rpm) at room temperature or overnight at 4 °C.
7. Wash the membrane 3 times with 1× PBS solution (100 rpm) for 5–10 min each time.

8. Incubate the membrane again for 1 h, this time with the second antibody.
9. Wash the membrane 3 times with 1× PBS solution (100 rpm) for 5–10 min each time.
12. Detect the proteins with the Western Blot Detection kit in a chemiluminescent system according to the manufacturer's instructions (Fig. 1).

4 Notes

1. Prepare stock solutions and keep them at 4 °C. Prepare protein extraction buffer the day of the experiment from the stock solutions. Be careful to make enough extraction buffer for all the samples. Generally, approximately 100 μL of buffer is used for ground tissue from an entire plate (with approximately 100 seedlings) that reaches the mark of 500 μL in a 1.5-mL tube.
2. Prepare stock solutions of 1 M DTT and 0.1 M PMSF. PMSF is a protease inhibitor and toxic, and not soluble in water;

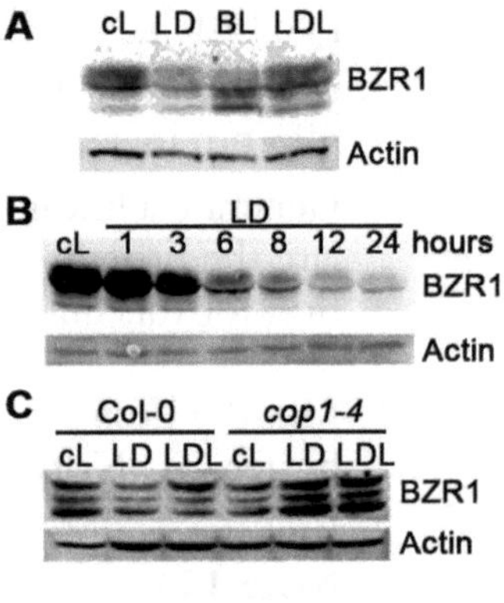

Fig. 1 Protein gel blots showing BZR1 protein abundance under different light regimes. (**a**) Seedlings overexpressing BZR1 under the cauliflower mosaic virus (CaMV) 35S promoter were subjected to the designated treatments. Continuous light (cL), light and transfer to darkness for 6 h (LD), brassinolide (BL) treatment, and transfer back to light after darkness (LDL). (**b**) BZR1 abundance in samples. The seedlings were transferred to darkness for 1, 3, 6, 8, 12, and 24 h before analysis. BZR1 protein was detected with the primary anti-hemagglutinin (HA) rabbit antibody and secondary goat anti-rabbit antibody. (**c**) Treatments of Col-0 wild type or *cop1-4* mutant seedlings. They were maintained in light (cL), transferred to dark for 6 h (LD), or put back 6 h in light after 6 h dark treatment (LDL). Native BZR1 abundance was detected using rabbit anti-BZR1 antibody [16] and secondary goat anti-rabbit antibody. The *top bands* represent phosphorylated forms of BZR1, whereas the *bottom band* the dephosphorylated form. In all the blots, the mouse anti-actin antibody (along with goat anti-mouse secondary antibody) was used as a loading control of total protein

hence, dissolve in isopropanol, ethanol, or methanol. Keep the stocks in small aliquots at −20 °C and avoid repetitive freeze and thaw, because the compounds can lose efficacy. Addition to the extraction buffer is optional, but recommended.

3. Dissolve Tris base and glycine together in water and add methanol at the end. If 10× transfer buffer is prepared, avoid adding methanol directly to the 10× buffer because it precipitates the compounds. Add methanol just to 1× working solution.
4. Sow the seeds separately from one another. Consider one plate with approximately 100 seedlings for each light treatment to have enough tissue.
5. To obtain consistent results, check the time at which the light treatments are done and maintain it consistently when repeating the experiments, because protein levels can also be altered at different hours.
6. Try resuspension just by tapping the tube and avoid vortexing samples, because it heats the sample and can damage the proteins. If vortex, do it briefly and place it on ice for a few seconds each time to cool down the sample.
7. If the supernatant is not clear and contains remains of tissues, centrifuge the sample for an additional minute.
8. The measure of protein concentration is determining to compare equal amounts of total protein in the final result. Therefore, it is important to avoid pipetting errors when mixing the same amounts of reagent and protein in all samples. It is also suggested to do the measurements in triplicates.
9. Absorbance of the set of standards is plotted as a function of the concentration to obtain the standard curve. Concentrations are on the x-axis and absorbance on y-axis. Unknown concentrations are calculated with the $x = (y - b)/m$ equation (where m is the slope and b the intercept) with the absorbance values. When using Excel, both the equation and R-square values are displayed as options after adding a linear trendline to the data.
10. The amount of protein to be loaded in a gel will depend on the protein levels. When working with transgenic plant lines that overexpress the protein of interest, 15 μg of total protein may be enough to clearly detect the protein. When endogenous proteins are to be detected, then 25 or 30 μg of total protein is recommended.
11. Avoid putting the samples on ice once mixed with the buffer, because it contains SDS and it could precipitate.
12. Transfer can be done at low voltage (20 V) overnight or for 1–2 h at 100 V, if a cooling system is used to maintain the buffer at 4 °C. High voltages increase the temperature of the buffer and may damage the samples.

Acknowledgments

This research was supported, in part, by grants from the Next-Generation BioGreen21 Program (PJ01104501), Cooperative Research Program for Agricultural Science and Technology Development (Project No. PJ01168501), Rural Development Administration, and the National Research Foundation of Korea (NRF) grant funded by the Korea government (MSIP) (No. 2015R1A2A1A10051668) Republic of Korea.

References

1. Chory J, Li J (1997) Gibberellins, brassinosteroids and light-regulated development. Plant Cell Environ 20:801–806
2. Wang Z-Y, Bai M-Y, Oh E et al (2012) Brassinosteroid signaling network and regulation of photomorphogenesis. Annu Rev Genet 46:701–724
3. Noguchi T, Fujioka S, Takatsuto S et al (1999) Arabidopsis *det2* is defective in the conversion of (24*R*)-24-methylcholest-4-en-3-one to (24*R*)-24-methyl-5α-cholestan-3-one in brassinosteroid biosynthesis. Plant Physiol 120:833–840
4. Szekeres M, Németh K, Koncz-Kálmán Z et al (1996) Brassinosteroids rescue the deficiency of CYP90, a cytochrome P450, controlling cell elongation and de-etiolation in Arabidopsis. Cell 85:171–182
5. Kim B, Jeong YJ, Corvalán C et al (2014) Darkness and *gulliver2/phyB* mutation decrease the abundance of phosphorylated BZR1 to activate brassinosteroid signaling in Arabidopsis. Plant J 77:737–747
6. Li Q-F, He J-X (2015) BZR1 interacts with HY5 to mediate brassinosteroid- and light-regulated cotyledon opening in *Arabidopsis* in darkness. Mol Plant 9:113–125
7. Oh E, Zhu J-Y, Wang Z-Y (2012) Interaction between BZR1 and PIF4 integrates brassinosteroid and environmental responses. Nat Cell Biol 14:802–809
8. He J-X, Gendron JM, Yang Y et al (2002) The GSK3-like kinase BIN2 phosphorylates and destabilizes BZR1, a positive regulator of the brassinosteroid signaling pathway in *Arabidopsis*. Proc Natl Acad Sci U S A 99:10185–10190
9. Luo X-M, Lin W-H, Zhu S et al (2010) Integration of light- and brassinosteroid-signaling pathways by a GATA transcription factor in *Arabidopsis*. Dev Cell 19:872–883
10. de Lucas M, Prat S (2014) PIFs get BRright: PHYTOCHROME INTERACTING FACTORs as integrators of light and hormonal signals. New Phytol 202:1126–1141
11. Gallagher S, Winston SE, Fuller SA et al (2008) Immunoblotting and immunodetection. Curr Protoc Immunol 83:1–28. (Chapter 8, Unit 8.10)
12. Shewry PR, Fido RJ (1996) Protein extraction from plant tissues. In: Doonan S (ed) Protein purification protocols, methods in molecular biology, vol 59. Humana Press, Totowa, pp 23–29
13. Conlon HE, Salter MG (2007) Plant protein extraction. In: Rosato E (ed) Circadian rhythms: methods and procotols, methods in molecular biology, vol 362. Humana Press, Totowa, pp 379–383
14. Cleland WW (1964) Dithiothreitol, a new protective reagent for SH groups. Biochemistry 3:480–482
15. Gallagher SR, Winston SE, Fuller SA et al (2001) Immunoblotting and immunodetection. Curr Protoc Cell Biol 52:1–28. (Chapter 6, Unit 6.2)
16. Jeong YJ, Corvalán C, Kwon S et al (2015) Analysis of anti-BZR1 antibody reveals the roles BES1 in maintaining the BZR1 levels in Arabidopsis. J Plant Biol 58:87–95

Chapter 5

Approaches to Study Light Effects on Brassinosteroid Sensitivity

Sandi Paulišić, Maria José Molina-Contreras, Irma Roig-Villanova, and Jaime F. Martínez-García

Abstract

Light perception and hormone signaling in plants are likely connected at multiple points. Light conditions, perceived by photoreceptors, control plant responses by altering hormone concentration, tissue sensitivity, or a combination of both. Whereas it is relatively straightforward to assess the light effects on hormone levels, hormone sensitivity is subjected to interpretation. In *Arabidopsis thaliana* seedlings, hypocotyl length is strongly affected by light conditions. As hypocotyl elongation also depends on brassinosteroids (BRs), assaying this response provides a valuable and easy way to measure the responsiveness of seedlings to BRs and the impact of light. We describe a simple protocol to evaluate the responsiveness of hypocotyls to commercial BRs and/or BR inhibitors under a range of light conditions. These assays can be used to establish whether light affects BR sensitivity or whether BRs affect light sensitivity. Overall, our protocol can be easily applied for deetiolation (under polychromatic or monochromatic light) and simulated shade treatments combined with BR treatments.

Key words Monochromatic light, Dark, Shade, Hypocotyl length, Deetiolation, Photoreceptors, Hormone sensitivity

1 Introduction

The development of young seedlings is determined by multiple environmental factors, of which light is a key signal. Light conditions are perceived by a battery of photoreceptors, such as the blue light-absorbing cryptochromes or the red/far-red light-absorbing phytochromes, that interpret these light signals and activate highly coordinated and regulated transcriptional networks. Young seedlings that germinate in the dark follow a developmental program known as skotomorphogenesis, characterized by elongated seedling stems (hypocotyls), unopened apical hooks, as well as closed cotyledons. Upon exposure to light, seedlings switch to a different developmental process, called photomorphogenesis, in which hypocotyl elongation is inhibited, the apical hook and the

Eugenia Russinova and Ana I. Caño-Delgado (eds.), *Brassinosteroids: Methods and Protocols*, Methods in Molecular Biology, vol. 1564, DOI 10.1007/978-1-4939-6813-8_5,

cotyledons open, and the seedlings become photosynthetically active. Exposure of emerging seedlings to light conditions that simulate vegetation proximity also promotes hypocotyl elongation and might inhibit cotyledon expansion [1, 2].

A central hub in the activation of the gene expression pattern required for the implementation of the corresponding developmental program is constituted by the PHYTOCHROME INTERACTING FACTORs (PIFs). These transcription factors are regulated by the circadian clock and temperature, and their stability and/or DNA-binding abilities are affected through interaction with phytochromes and cryptochromes [3, 4] that initiate and modulate a regulatory transcriptional network.

Brassinosteroids (BRs), like auxins and gibberellins, are a group of plant hormones that generally promote cell elongation and expansion. Overall, there are multiple contact points between light and hormone signaling that likely link light perception and the actual changes in plant growth and physiology. Specifically, BRs play an important role in photomorphogenesis and act as negative regulators of hypocotyl elongation and positive regulators of cotyledon expansion [5]. In *Arabidopsis thaliana*, the elongation of hypocotyls in the dark, under monochromatic light, or in response to plant proximity or shade requires BRs [6]. The light control of hormone-triggered responses might be exerted by concentration changes, alterations in tissue sensitivity, or a combination of both [7, 8]. Whereas measuring hormone levels is straightforward with the adequate analytical approaches, measuring hormone sensitivity is subjected to interpretation and is deduced from the responsiveness of specific tissues to exogenous hormones and/or plant growth regulators. It has been suggested that both hormone levels and sensitivity changes are required to promote hypocotyl elongation [7]. Furthermore, measuring the length of *Arabidopsis* hypocotyls is a valuable and simple way to measure the responsiveness of seedlings to light and BRs [5, 9], as well as to establish whether light affects BR sensitivity [8, 10, 11] or BRs affect light sensitivity [12].

This chapter provides a simple protocol for measuring the responsiveness of hypocotyls to different concentrations of commercial BRs and/or their inhibitors under various light conditions or regimes. We describe the use of different light conditions that can be easily tested to perform (1) deetiolation experiments using increasing amounts of polychromatic or monochromatic light, (2) simulated shade treatments, and (3) combinations of (1) and (2) with hormonal treatments.

2 Materials

2.1 Medium

1. Growth medium: half-strength Murashige and Skoog (½MS) mineral salts, 0.8 % (w/v) agar, and no sucrose (*see* **Note 1**). Mix 2.15 g MS salts and 0.8 g 2-(*N*-morpholino)ethanesulfonic acid (MES) and 900 mL water. Stir to dissolve. Adjust pH to 5.7 with KOH. Adjust volume with water to 1 L. Add 8 g Bacto agar. Sterilize by autoclaving.

2.2 Sterilization Solutions and Equipment

1. Imbibition solution: 0.1 % (v/v) polyoxyethylene sorbitan monolaureate (Tween 20).
2. Sterilization solution: 0.1 % (v/v) Tween 20, 10 % (v/v) sodium hypochlorite.
3. Sterile milliQ water.
4. Microcentrifuge tubes (1.5–2 mL).
5. Water bath.
6. Pasteur pipettes or standard blue pipette tips (200–1000 μL).
7. Sterile nylon mesh (100 μM, Nitex mesh).
8. Petri dishes (90 mm diameter and 16 mm height).
9. Porous surgical tape (e.g., tape 1.25 cm × 9.1 m).

2.3 Light Sources and Treatments

1. Aluminum foil for dark treatments (*see* **Note 2**).
2. Monochromatic blue (B), red (R) and far-red (FR) light provided by light-emitted diodes (LEDs) (for instance, GreenPower LED module HF blue, deep red, and far-red from Philips).
3. White light (W) provided by cool-white fluorescent bulbs (for instance, Philips TL-D 36w/840).
4. Filter paper (60 g/m^2) and grey filter to obtain different light intensities by placing several layers of neutral filters between the light sources and the seed-containing plates (*see* **Note 3**).
5. W supplemented with monochromatic FR LEDs to simulate shade treatments.
6. Spectroradiometer to measure fluence rates (such as a Spectrosense2 meter associated with a 4-channel sensor; Skye Instruments Ltd.) that quantifies photosynthetically active radiation (PAR) (400–700 nm) and 10 nm windows in the B (463–472 nm), R (664–673 nm), and FR (725–734 nm) regions.

2.4 Plant Growth Regulators

1. Prepare the biologically active BR, epibrassinolide (eBL), (CAS 78821-43-9) as a 5 mM stock solution in ethanol 50 % (v/v) and store it at −20 °C until use. It can be added to the MS

medium in a range of concentrations (from 0.01 to 10^{-6} M) and poured into plates.

2. Prepare the potent BR synthesis inhibitor brassinazole (BRZ) as a 10 mM stock solution in dimethoxysulfoxide (DMSO) and store it at −20 °C until use. It can be added to the MS medium in a range of concentrations (from 10 to 1000 nM) and poured into plates.
3. Prepare propiconazole (PCZ), the active ingredient of a broad-spectrum fungicide that acts as a BR synthesis inhibitor in both *Arabidopsis* and maize (*Zea mays*) seedlings [13] as a 5 mM stock solution in water and store it at −20 °C until use.

2.5 Software for Hypocotyl Measurements

1. ImageJ software (http://rsb.info.nih.gov/) to measure the hypocotyl lengths of the seedlings.

3 Methods

3.1 Seed Surface Sterilization

1. Aliquot the amount of seeds needed for the experiment in microcentrifuge tubes (1.5–2 mL) for sterilization. The weight of 100 *Arabidopsis* (accession Columbia-0) wild-type dry seeds is approximately 0.002 g (*see* **Notes 4–6**).
2. Add 1 mL of imbibition solution for at least 30 min.
3. After removal of the imbibition solution, add 1 mL of sterilization solution for 10 min (not more).
4. Remove the sterilization solution and wash the seeds with 1 mL of sterile water (5 times) before sowing. Different sowing methods will depend on the experimental treatment you plan to perform (*see* Subheading 3.3).

3.2 Preparation of Plates for Sowing Seeds

For long-term treatments and hypocotyl experiments, growth regulators are added to the 1/2MS medium after cooling it to 55 °C in a water bath (*see* **Notes 7–9**). *Arabidopsis* seedlings are grown in regular petri dishes containing approximately 25 mL of sterile growth medium (1/2MS).

3.3 Seed Sowing

1. For long-term treatments and hypocotyl experiments, sow the seeds directly on the solid medium one by one using a micropipette or a Pasteur pipette (*see* **Notes 10–14**).
2. For short hormone treatments and/or analysis of gene expression, sow the seeds on a sterile nylon mesh (*see* **Note 15**).
3. After sowing, seal the plates with a porous surgical tape (*see* **Note 16**).

3.4 Light Treatments of Seedlings for Hypocotyl Elongation Analysis

1. For stratification, keep the plates containing the sown seeds in the dark at 4 °C for 2–5 days. This stratification treatment breaks dormancy and synchronizes the germination of the seeds.
2. For the deetiolation experiments, induce the germination by a brief (0.5–3 h) W treatment. Then, wrap the plates with at least two layers of aluminum foil (*see* **Note 2**) and keep them in the dark at 22 °C for up to 24 h from the moment germination was induced (*see* **Note 17**). Thereafter, keep one plate in darkness at 22 °C for the whole treatment (this is the dark control treatment in which seedlings elongate to their maximum capacity) and transfer the remainder of the plates to the different polychromatic (W) or monochromatic (B, R, or FR) light conditions (Fig. 1). We measure the hypocotyl length on day 4 from germination.
3. For the simulated shade treatments, after stratification, put plates to germinate in continuous W at 22 °C. On day 2, maintain half the plates under W and transfer the other half to W+FR. We measure the hypocotyl length on day 7 from germination (after 5 days of differential light/shade treatment) (Fig. 2) (*see* **Note 17**).

3.5 Measurement of Hypocotyls

1. To measure hypocotyl length, lay seedlings flat on the agar plates.
2. Take digital pictures of the plates. Include a reference scale (*see* **Note 18**).
3. Measure hypocotyl length by analyzing the digital images of the seedlings with Image J or similar software (*see* Subheading 2.5).

4 Notes

1. Addition of sucrose to the medium interferes with the hypocotyl phenotypes in seedlings, therefore avoid it.
2. For the deetiolation experiments (Fig. 1), use at least a double layer of aluminum foil when keeping the plates in the dark (i.e., in the initial 24 h of darkness before exposing them to light or in the dark-grown control seedlings). This additional precaution will avoid even small amounts of light that might reach the seedlings through small pores of the aluminum foil, hence impairing the experiment.
3. We recommend preparing boxes with the filters (*see* Fig. 1d) to obtain more reproducible results instead of putting filters just on top of the plates.

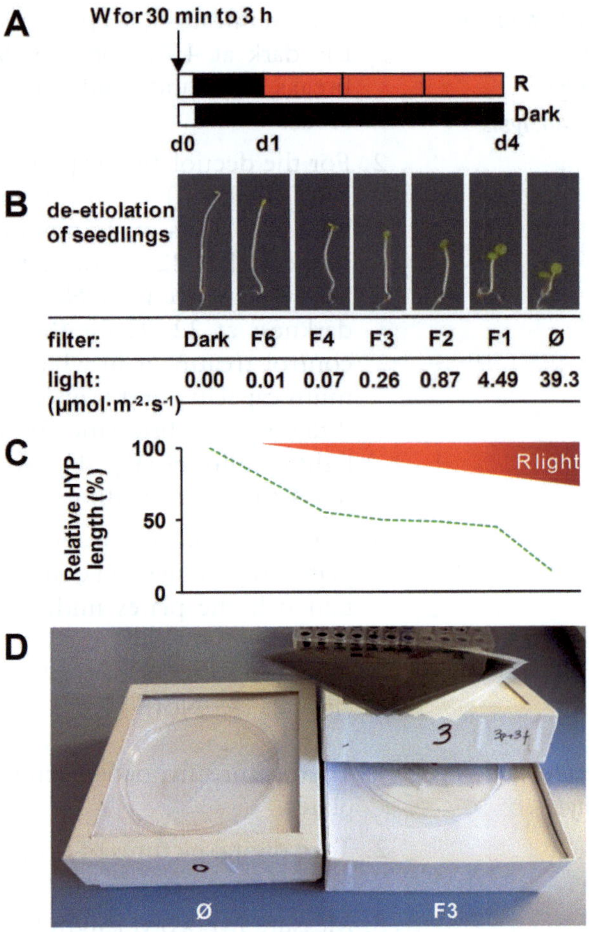

Fig. 1 Deetiolation of *Arabidopsis* seedlings under increasing amounts of light. (**a**) After induction of germination with a pulse of 2 h of white light (W), transfer of plates to the dark for 1 day followed by 3 days in monochromatic red light (R). As a control, a plate is kept in the dark. (**b**) Seedlings grown in the dark (*left*) or in increasing intensities of R (to the *right*). Dark-grown seedlings have long hypocotyls (HYP) and unopened apical hooks. R intensity is controlled by covering the plates with filters (F6-F1) that provide a range from 0.01 (F6) to 39.3 μmol·m^{-2}·s^{-1} without filter (Ø). (**c**) PhyB perception of R allowing transition from skotomorphogenesis to photomorphogenesis and decreasing hypocotyl lengths with increasing R intensity. (**d**) Example of boxes providing different light intensities. Box 3 (*right*) has three layers of grey filters and three layers of filter papers and corresponds to the F3 data (B). Box without filters (*left*) used for growing the seedlings in full R intensity, corresponds to Ø data (B)

4. When working with seedling phenotypes, always include the reference wild type together with the mutants of interest.
5. It is particularly important to compare seedlings from seeds originating from mother plants grown at the same time and under the same conditions, because any changes in seed age or

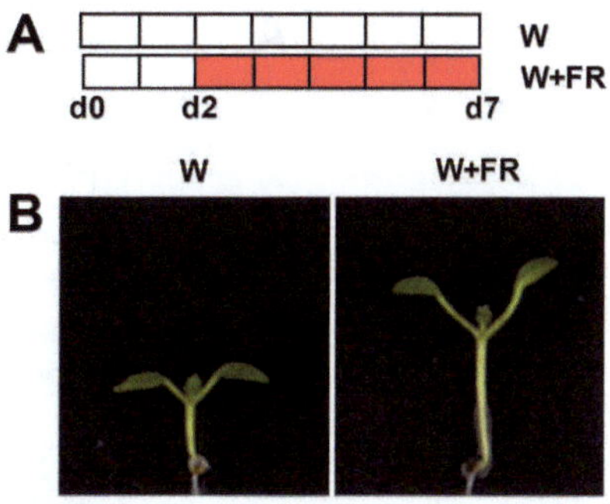

Fig. 2 Responses of *Arabidopsis* seedlings to simulated shade treatments. (**a**) *Arabidopsis* seedlings grown for 7 days (d) in continuous white light (W) or 2 days in W and 5 days in W supplemented with far-red light (W + FR). (**b**) Seedlings grown as indicated in a. Seedlings grown in W + FR show elongated hypocotyls, typical for shade avoidance

state could affect germination and, thus, influence the hypocotyl phenotypes.

6. For the hypocotyl measurements, use at least 20 seedlings for each data point and/or genotype. Repeat the experiments at least 3 times to help you carry out statistical analyses of the elongation data.
7. For experiments with hormones and inhibitors (Fig. 3), we recommend preparing serial dilutions of stocks for the different concentrations to test. The same stock can be used for different experiments as far as they are stored at −20 °C.
8. It is important to carefully homogenize the 1/2MS growth medium when hormones and/or inhibitors are added.
9. We normally use petri dishes of 90 mm diameter and 16 mm height (the same ones used to grow bacteria). Other sizes can also be employed, depending on the number of lines to compare.
10. After bleach sterilization and water washes, add 1 mL of sterile water. Seeds can be sown individually on top of the media with an automatic pipette and blue tips. Pull a small volume of water containing approximately 20 seeds. They will easily fall one by one just by touching the medium surface.
11. Some seed batches are too large to get inside the blue tips. In these cases, sterile Pasteur pipettes coupled with a rubber sucker can be used.
12. For light experiments, it is important to distribute seeds separately to avoid negative effects of elongation because of plant proximity or touching.
13. We normally sow about 100 seeds per petri dish, one by one. A plate can be divided in a maximum of four sections; a geno-

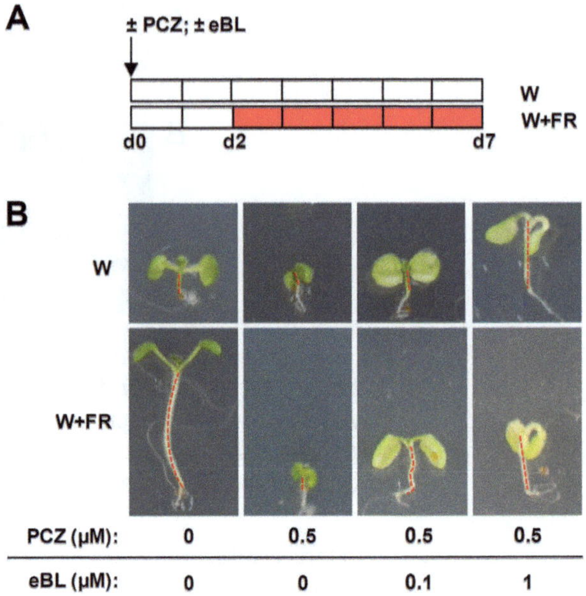

Fig. 3 Effect of simulated shade on the hypocotyl response to hormone application. (**a**) *Arabidopsis* Col-0 wild-type seedlings germinated and grown as indicated in Fig. 1a. Medium is supplemented from the day of sowing with propiconazole (PCZ) and with or without 24-epibrassinolide (eBL). (**b**) Seedlings grown as indicated (a) with the amounts of PCZ and eBL indicated below the images. *Red-dashed lines* mark hypocotyls of the representative seedlings grown as described

type is sown in each section (approximately 25 seeds per genotype).

14. Use a paper with a grid to sow regularly spaced seeds. Make sure any remaining water is completely evaporated before sealing the plates to start the stratification.
15. If you want to harvest material to analyze gene expression or protein extraction, sow the seeds on top of a nylon mesh (100 μM, Nitex Mesh) previously cut and sterilized. It will be easier and faster to harvest the plant material for further processing.
16. Usage of nonporous sealer (such as Parafilm or plastic wrap) is not recommended, because we have observed that hypocotyl elongation is clearly affected in such a closed atmosphere.
17. Be strict with the timing, especially for short light or hormone treatments. Perform the experiments always at the same time of the day, because circadian rhythms can influence the experiments. For instance, when plates are taken out from stratification to induce germination at 11 a.m., transfer half of the plates into W+FR at day 2 at 11 a.m. and end the experiment and take pictures on day 7 also at 11 a.m.

18. Remember to add a scale when taking the digital pictures of the hypocotyls, because it will be needed for measurement of the hypocotyl lengths.

Acknowledgments

SP received a predoctoral fellowship from the Catalan Agència de Gestió d'Ajuts Universitaris i de Recerca (AGAUR, FI-DGR 2015). MJM-C received a predoctoral FPI fellowship from the Spanish Ministry of Economy and Competivity (MINECO). IR-V received a Beatriu de Pinós contract from AGAUR. Our research is supported by grants from the MINECO-FEDER (BIO2011-23489, BIO2014-59895-P) and AGAUR (2014-SGR447 and Xarba) to JFM-G.

References

1. Casal JJ (2013) Photoreceptor signaling networks in plant responses to shade. Annu Rev Plant Biol 64:403–427
2. Yu X, Liu H, Klejnot J et al (2010) The cryptochrome blue light receptors. The Arabidopsis Book 8:e0135
3. Leivar P, Quail PH (2011) PIFs: pivotal components in a cellular signaling hub. Trends Plant Sci 16:19–28
4. Pedmale UV, Huang S-sC, Zander M et al (2016) Cryptochromes interact directly with PIFs to control plant growth in limiting blue light. Cell 164:233–245
5. Turk EM, Fujioka S, Seto H et al (2005) *BAS1* and *SOB7* act redundantly to modulate Arabidopsis photomorphogenesis via unique brassinosteroid inactivation mechanisms. Plant J 42:23–34
6. Wang Z-Y, Bai M-Y, Oh E et al (2012) Brassinosteroid signaling network and regulation of photomorphogenesis. Annu Rev Genet 46:701–724
7. Alabadí D, Blázquez MA, Carbonell J et al (2009) Instructive roles for hormones in plant development. Int J Dev Biol 53:1597–1608
8. Bou-Torrent J, Galstyan A, Gallemí M et al (2014) Plant proximity perception dynamically modulates hormone levels and sensitivity in *Arabidopsis*. J Exp Bot 65:2937–2947
9. Koornneef M, van der Veen JH (1980) Induction and analysis of gibberellin sensitive mutants in *Arabidopsis thaliana* (L.) Heynh. Theor Appl Genet 58:257–263
10. Turk EM, Fujioka S, Seto H et al (2003) CYP72B1 inactivates brassinosteroid hormones: an intersection between photomorphogenesis and plant steroid signal transduction. Plant Physiol 133:1643–1653
11. Luo X-M, Lin W-H, Zhu S et al (2010) Integration of light- and brassinosteroid-signaling pathways by a GATA transcription factor in *Arabidopsis*. Dev Cell 19:872–883
12. Oh E, Zhu J-Y, Wang Z-Y (2012) Interaction between BZR1 and PIF4 integrates brassinosteroid and environmental responses. Nat Cell Biol 14:802–809
13. Hartwig T, Corvalan C, Best NB et al (2012) Propiconazole is a specific and accessible brassinosteroid (BR) biosynthesis inhibitor for *Arabidopsis* and maize. PLoS One 7:e36625

Chapter 6

A Technical Framework for Studying the Signaling Nexus of Brassinosteroids and Immunity

Rosa Lozano-Durán and Youssef Belkhadir

Abstract

Pathway cross-communication cannot be simply tackled by studying isolated signaling systems. Yet understanding how signal transduction pathways attenuate or reinforce each other in vivo is a challenging task. In plants, biosynthesis and signaling of brassinosteroids (BRs) finely regulate growth and defense programs through a complex array of mechanistic and physiological interactions. Conversely, induction of defenses also impacts on the BR biosynthesis at the transcriptional level. In this chapter, we present an experimental framework to study the physiological connection between BR-controlled growth and defenses. We focus on the signaling pathways regulated by the two archetypal cell surface receptors, BRASSINOSTEROID INSENSITIVE1 (BRI1) and FLAGELLIN-SENSITIVE2 (FLS2), to illustrate the signaling nexus of BRs and plant immunity. In *Arabidopsis thaliana*, these pathways provide one of the very few systems in which the tools and mechanistic details exist to study cross talk at the molecular level.

Key words FLAGELLIN-SENSITIVE 2 (FLS2), BRASSINOSTEROID INSENSITIVE 1 (BRI1), BRI1-ASSOCIATED RECEPTOR KINASE 1 (BAK1), Immunity, Growth, Cross talk, Receptor interaction, Reactive oxygen species (ROS), MAPKs activation, Callose deposition, Regulation of gene expression, Seedling growth inhibition

1 Introduction

Plants need to grow while protecting themselves against various types of pathogens [1, 2]. As defense and growth programs are both costly, the plant successful colonization of resource-scarce environments often leads to a dilemma in which trade-off decisions have to be made [1, 2]. Plants use signaling cross talk between pathways that control growth and immunity to resolve their resource allocation problem in a timely and efficient manner [1, 2]. Evidence is accumulating that plant developmental programs are layered atop inducible defense systems through direct molecular convergence links [3]. In the past five years, fast-paced research in *Arabidopsis thaliana* has uncovered a multilevel cross talk between brassinosteroid (BR) signaling and plant immunity.

Eugenia Russinova and Ana I. Caño-Delgado (eds.), *Brassinosteroids: Methods and Protocols*, Methods in Molecular Biology, vol. 1564, DOI 10.1007/978-1-4939-6813-8_6,

BR signaling provides a coordinated input in the first layer of plant defenses [4]. Briefly, this layer relies on the detection of conserved pathogen-associated molecular patterns (PAMPs) by pattern recognition receptors (PRRs) at the cell surface [5]. PAMP binding to their cognate PRRs triggers a signaling cascade that ultimately leads to the onset of PAMP-triggered immunity (PTI) [6]. PTI can efficiently repel most pathogens, but its activation usually takes a toll on plant growth [1]. To date, the most extensively studied plant PRR is the *Arabidopsis* leucine-rich-repeat receptor kinase (LRR RK) FLAGELLING SENSING2 (FLS2) that recognizes directly bacterial flagellin (or its active epitope, the peptide flg22), the main component of the flagellum [5]. Flg22 detection by FLS2 elicits a signaling cascade that is not fully understood [1]. Direct binding of flg22 to FLS2 induces its association with the coreceptor BRI1-ASSOCIATED KINASE1 (BAK1) [7]. This receptor/coreceptor interaction is a requisite for the activation of the FLS2 pathway [1] (Fig. 1). Recruitment of BAK1 to FLS2 is then followed by a continuum of molecular events that occur in a time-dependent manner [1, 2]. Rapid FLS2 responses happen within seconds, minutes, or hours and are commonly monitored by output measurements, such as production of reactive oxygen species (ROS), subsequent activation of mitogen-associated kinases (MAPKs), and changes in target gene expression [1, 2] (Fig. 1). Long-term FLS2 responses take place over hours to days and can be monitored by output quantifications, such as callose deposition at the cell wall or seedling growth inhibition (SGI) (Fig. 1) [1, 2]. SGI is thought to represent an extrapolated growth inhibition process that normally occurs less intensely upon perception of physiological doses of flg22.

The overall architecture of the flg22 and BR response pathways is strikingly similar (Fig. 1) [2]. First, the ligands activate their respective receptors by direct binding in the extracellular space [2]. Second, ligand perception triggers BAK1 recruitment to the respective extracellular domains of the receptors (Fig. 1) [2]. Finally, the activation of the signaling pathways converges into massive transcriptional responses that govern antagonistic processes, such as cellular elongation (BR) and immunity (flg22) [1, 2, 4]. The common principles for receptor activation, together with the opposing physiological outcomes of both pathways, led researchers to speculate that a nexus could exist between BR and flg22 signaling. This cross talk between pathways has been confirmed and

Fig. 1 (continued) Standard range of concentrations and approximate treatment times are given as an indication. Details for approaches involving exogenous application of BL and/or flg22 are indicated on *top* of the arrows. Details for approaches involving steady-state manipulation of BR signaling or biosynthesis (gain-of-function approach) are given at the *bottom* of the arrows

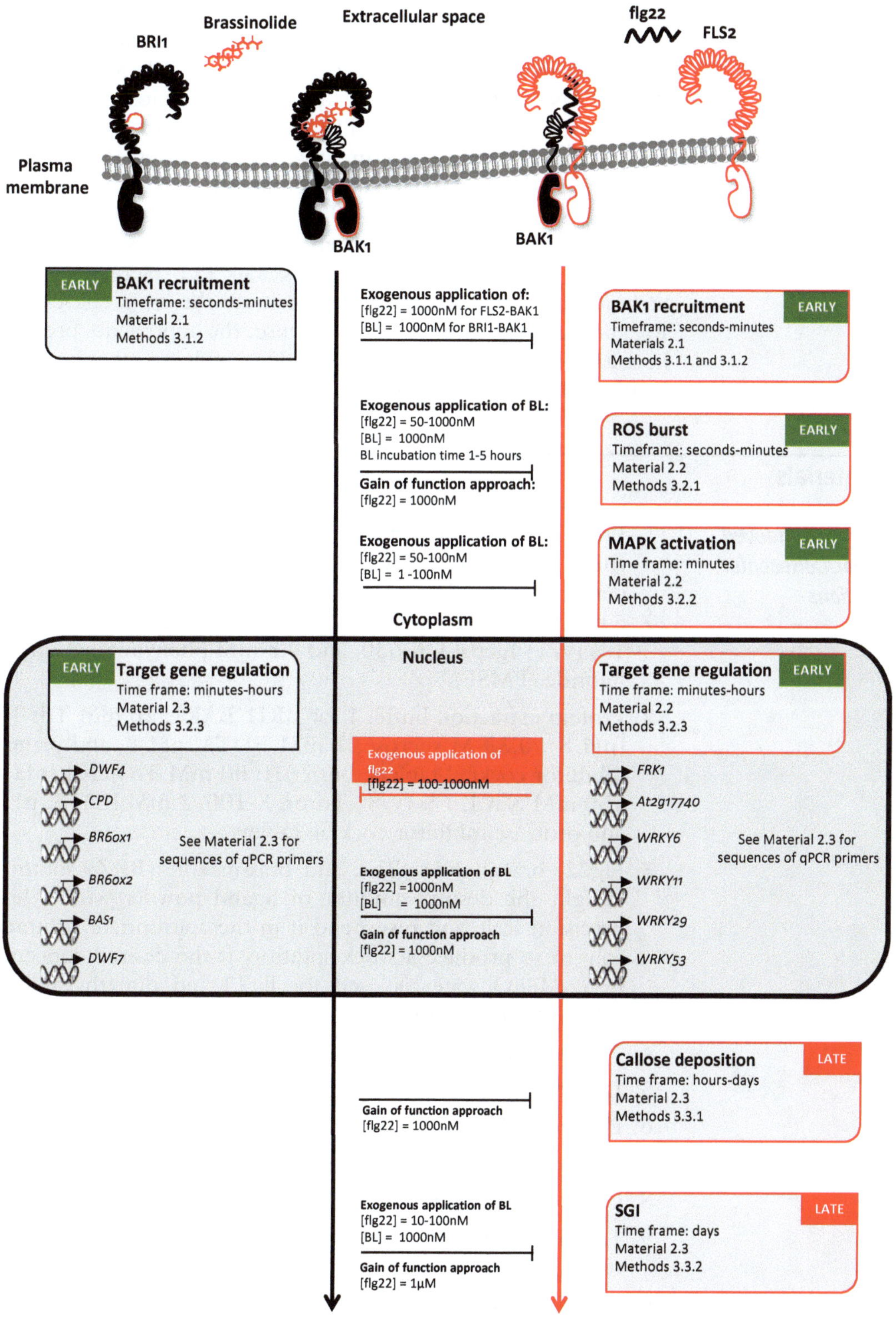

Fig. 1 Response outputs from the interaction between BR and flg22 signaling. Emphasis is put on the inhibitory relationship between the two pathways. *Blunt-ended arrows* mark inhibitory events and *regular arrows* positive regulatory events. For each readout, a reference to a specific material and method section is assigned.

shown to take place at different levels. Elevated BR signaling suppresses flg22 responses and this inhibitory effect largely depends on the activity of the transcription factor BRASSINAZOLE-RESISTANT1 (BRZ1) [8–11]. In return, BR-biosynthetic genes are down-regulated during PTI [12], probably to reduce the BR signaling output and, consequently, alleviate the steady-state inhibition of defenses by BRs. Although several direct physical intersection points linking BR and PTI signaling have been identified at the receptor level [13, 14], current models favor a signaling nexus driven by direct transcriptional cross talk [1]. Here, we present technical procedures used to demonstrate the cross talk between BR and flg22 signaling. In a wide sense, the approaches presented here can provide information for studies involving other hormone-controlled or PAMP signaling pathways.

2 Materials

2.1 Ligand-Induced Receptor/CoReceptor Interactions

1. Protein extraction buffer A for FLS2-BAK1: 50 mM Tris–HCl (pH 7.5), 150 mM NaCl, 10 % (v/v) glycerol, 2 mM ethylenediaminetetraacetic acid (EDTA) and water. Before use, add 5 mM dithiothreitol (DTT), protease inhibitor cocktail tablets, 1 % (v/v) Igepal CA-630, and 0.5 mM phenylmethylsulfonylfluoride (PMSF).
2. Protein extraction buffer I for BRI1-BAK1: 20 mM Tris-HCl (pH 8), 0.33 M sucrose, 1 mM EDTA, pH 8, and protease inhibitor cocktail tablets; buffer II: 50 mM Tris-HCl (pH 8), 150 mM NaCl, 1 % (v/v) Triton X-100, 1 mM EDTA, pH 8, and protease inhibitor cocktail tablets.
3. Flg22, brassinolide (BL), and brassinazole (BRZ) solutions. Weight the desired amount of ligand powder with a high-precision scale and resuspend it in the appropriate volume of solvent to produce a stock solution at the desired concentration. MilliQ water is used for flg22 and dimethylsulfoxide (DMSO) or ethanol for BL and BRZ.
4. Mortars, pestles, and spatulas.
5. Liquid nitrogen.
6. Polytron (Kinematica or equivalent).
7. Glass douncer (Kontes Glass Company or equivalent).
8. Cooled centrifuge.
9. 1.5-mL, 5-mL tubes (Eppendorf or equivalent), and 50-mL tubes.
10. Rotating wheel.
11. Vacuum pump.

12. Half-strength Murashige and Skoog (1/2 MS) medium containing 1 % (w/v) sucrose and 0.8 % (w/v) agar.
13. Sterile petri dishes (90 mm diameter).
14. Poly-prep chromatography column (Bio-Rad).
15. Beads (Trueblot beads, eBiosicence).
16. 5× loading buffer:1.5 g of sodium dodecyl sulfate (SDS), 3.75 mL 1 M Tris–HCl (pH 6.8), 0.015 g of bromophenol blue, 1.16 g DTT, and 7.5 mL of glycerol. Mix thoroughly with stirring. Add water until 15 mL total volume. Make 1-mL aliquots and store at −20 °C.
17. Novex™ 8 % Tris-glycine Mini Protein Gels (or equivalent).
18. Miracloth paper.

2.2 Early Readouts: ROS Burst, MAPK Activation, and Target Gene Expression

1. ROS detection solution: 100 μM Luminol, 10 μg/mL horseradish peroxidase (HRP), 100 nM flg22 peptide, and water.
2. Multiplate reader/luminometer (Tecan, Varioskan, or Biotek Synergy, or equivalent).
3. Lacus buffer: 50 mM Tris–HCl (pH 7.5), 10 mM $MgCl_2$, 15 mM ethylene glycol-bis (β-aminoethyl ether) *N,N,N′,N′*-tetraacetic acid (EGTA), 100 mM NaCl, 1 mM sodium fluoride (NaF), 1 mM sodium molybdate (NaMo), 0.5 mM Na_3VO_4 (activated), 30 mM β-glycero-phosphate, 0.1 % (v/v) Igepal CA-630, and water. Before use, add: 0.5 mM PMSF, protease inhibitor cocktail tablets, 100 nM calyculin A, and 2 mM DTT.
4. Phospho-p44/42 MAPK (Erk1/2; Thr202/Tyr204) rabbit monoclonal antibody.
5. RNA isolation kit: Spectrum Plant Total RNA Kit (Sigma-Aldrich or equivalent). For a custom-made protocol, see [15].
6. cDNA synthesis reaction: First-Strand cDNA Synthesis Kit (Fermentas) or SuperScript III RNA transcriptase (Invitrogen).
7. MyIQ Single Color Real-Time PCR Detection System (Bio-Rad) or Bio-Rad CFX96 real-time system or equivalent.
8. Gene specific primers:
   FRK1: Forward, 5′-ATCTTCGCTTGGAGCTTCTC-3′
   Reverse, 5′-TGCAGCGCAAGGACTAGAG-3′
   At2g17740: Forward, 5′-TGCTCCATCTCTCTTTGTGC-3′
   Reverse, 5′-ATGCGTTGCTGAAGAAGAGG-3′
   WRK11: Forward, 5′-ATGTCCAGCGAGGAAACACGT-3′
   Reverse, 5′-TATTCTCCTGCATCGCGGATT-3′
   WRKY6: Forward, 5′-CATATTACCGCTGCACGATGG-3′
   Reverse, 5′-GGCAACGGATGGTTATGGTTT-3′
   WRKY29: Forward, 5′-TTCGTTTTGCCTACCGATGG-3′
   Reverse, 5′-CGAGCTTGTGAGGATCGTTTG-3′
   WRK53: Forward, 5′-AAATCCCGGCAGTGTTCCA-3′

Reverse, 5′-TCTTGGCGATGATGACTCTCG-3′
DWF4: Forward, 5′-CATAAAGCTCTTCAGTCACGA-3′
Reverse, 5′-CGTCTGTTCTTTGTTTCCTAA-3′ (ref. 16)
CPD: Forward, 5′-CCCAAACCACTTCAAAGATGCT-3′
Reverse, 5′-GGGCCTGTCGTTACCGAGTT-3′
BR6ox1: Forward, 5′-TCAGTGGAATGAGCTTGAGG-3′
Reverse, 5′-AGTTCCCAGCGATTTGTGTC-3′ (ref. 17)
BR6ox2: Forward, 5′-GGAGGTGGAGTTAGGCTTTGC CC-3′
Reverse, 5′TCTTCTCCATTCTCTTCCCATCTAT-3′ (ref. 18)
BAS1: Forward, 5′-CATGTGGCACGGAGCTTCTA-3′
Reverse, 5′-CCGAGCTGGATTGAATTCGT-3′ (ref. 19)
DWF7: Forward, 5′-TTGCAAATGTTTGTGGCAAT-3′
Reverse, 5′-GATGGGTGGCATGGAGATAC-3′

9. Control primers:
   ACTIN 2/8: Forward, 5′-TCTTGTTCCAGCCCTCGTTT-3′
   Reverse, 5′-TCTCGTGGATTCCAGCAGCT-3′
   18S rRNA: Forward, 5′- TATAGGACTCCGCTGGCA CC-3′
   Reverse, 5′-CCCGGAACCCAAAAACTTTG-3′
   U-box (At5g15400): Forward 5′-TGCGCTGCCAGATAATA CACTATT-3′
   Reverse, 5′-TGCTGCCCAACATCAGGTT-3′ (ref. 20)
10. Cork borer size number 1 (3.8 mm diameter).
11. 1/2 MS medium (*see* Subheading 2.1, **step 12**).
12. Sterile petri dishes (90 mm diameter).
13. 24- and 96-well plates.
14. 5× loading buffer (*see* Subheading 2.1, **step 16**).
15. Novex™ 8 or 10 % Tris–Glycine Mini Protein Gels.
16. Eppendorf micropestles (or equivalent) for 1.5-mL tubes.

2.3 Late Readouts: Callose Deposition and Seedling Growth Inhibition (SGI)

1. Staining solution: 150 mM K_2HPO_4, pH 9.5, and 0.01 % (w/v) aniline blue.
2. Fixing solution: ethanol–acetic acid 3:1 (v/v).
3. 50 % (v/v) and 70 % (v/v) water–ethanol solutions.
4. Flg22 solution (*see* Subheading 2.1, **step 3**).
5. Vacuum pump.
6. Microscope Leica DM5000B or equivalent.
7. 50 % (v/v) glycerol solution.
8. MilliQ Water.
9. 0.5-mL Eppendorf tubes.
10. 6-well plates.
11. Liquid and solid 1/2 MS medium (*see* Subheading 2.1, **step 12**).

12. High-precision balance scale.
13. Sterile petri dishes (90 mm diameter).
14. UV microscope.

3 Methods

3.1 Ligand-Induced Receptor/Coreceptor Interactions

3.1.1 FLS2-BAK1 Interactions

1. Put sterile *Arabidopsis* seeds on plates containing the growth medium (*see* **Note 1**).
2. Stratify the seeds by keeping the plates in the dark at 4 °C for at least 2 days.
3. Transfer to light and let seedlings grow for 5 days under long-day conditions.
4. Transfer seedlings to liquid 1/2 MS medium containing 1 % (w/v) sucrose. Single Eppendorf tubes or multi-well plates can be used; 24-well plates allow to grow two seedlings per well. Per condition, 48 seedlings are enough.
5. Let the plants grow for another 7 days under long-day conditions.
6. The day before treatment and sampling, replace the liquid 1/2 MS medium with 500 μL of fresh liquid 1/2 MS medium.
7. Add 500 μL of solution containing flg22 and/or BL. Use the respective solvents as negative controls (*see* **Notes 2** and **3**). Wait for the desired time, blot dry, and freeze in liquid nitrogen. A 5-min flg22 treatment is enough to detect FLS2-BAK1 interactions.
8. Dry seedlings carefully and grind in liquid nitrogen with pestle in a prechilled mortar to obtain a fine green powder.
9. Transfer the finely ground tissue into ice-cooled 50-mL tubes.
10. Add 2 mL of protein extraction buffer A per 48 seedlings. Mix thoroughly and keep on ice.
11. Centrifuge at 4°C at more than 12,000 × *g* for 15 min.
12. Transfer the supernatant to a new tube and repeat the centrifugation step.
13. Filter the supernatant through a poly-prep chromatography column. Keep the flow-through on ice.
14. For each sample, equilibrate 40 μL of beads slurry with 1 mL of protein extraction buffer A.
15. Centrifuge at 500 × *g* for 30 s and discard the supernatant.
16. To the remaining 20 μL of beads, add 20 μL of extraction buffer A and 15 μL of anti-FLS2 or anti-BAK1 antibody (*see* **Note 4**).
17. Mix gently and keep the tubes on ice.

18. Add 1.5 mL of the protein extract to the beads, but before, take an aliquot of the total protein extract to use as input; for instance, take 40 μL of protein extract, add 10 μL of 5× loading buffer, and load 30 μL on the gel.
19. Incubate on a rotating wheel at 4 °C for 2–4 h.
20. Centrifuge at 500 × *g* for 30 s.
21. Discard the supernatant.
22. Resuspend the beads in 1 mL of protein extraction buffer A.
23. Repeat the washing steps at least 3 times.
24. Add 40 μL of 5× protein loading buffer to the beads.
25. Heat at 70 °C for 20 min.
26. Centrifuge at 500 × *g* for 1 min and load the supernatant on protein gels for western blotting (*see* **Note 5**).

3.1.2 BRI1-BAK1 Interactions

1. Put sterile *Arabidopsis* seeds on plates containing the growth medium.
2. Stratify seeds in the dark at 4 °C for at least 2 days.
3. Transfer to light and let seedlings grow for 2 weeks under short-day conditions.
4. Harvest and infiltrate 3 g of seedlings with a flg22 and/or BL solution by vacuum infiltration during 5 min. Pretreatments with BRZ help getting rid of the basal BAK1-BRI1 interaction, but they are not necessary. Make sure all seedlings are immersed in the treatment solution.
5. Dry seedlings carefully and grind in liquid nitrogen with a pestle in a prechilled mortar to obtain a fine green powder.
6. Transfer 1 g of powder in 5-mL Eppendorf tubes containing 2 mL of prechilled buffer I. Prior to use, add the desired ligands to buffer I (i.e., flg22/BL).
7. Homogenize by alternate rounds of Polytron.
8. Use Miracloth paper to filtrate 1.5 mL of the supernatant into a fresh 1.5-mL Eppendorf tube.
9. Centrifuge at 5,000 × *g* at 4 °C for 10 min or more until the supernatants are clear.
10. Discard the supernatants and gently resuspend the pellets in 0.5 mL of prechilled buffer II with a glass douncer. Prior to use, add the desired ligands to the buffer (i.e., flg22/BL).
11. Incubate for 30 min on a rotating wheel at 4 °C to solubilize the microsomal fraction.
12. Centrifuge for 10 min at 20,000 × *g* at 4 °C to clear the remaining debris.

13. Quantify the total protein concentrations by using absorbance at 280 nm.
14. Proceed as described (Subheading 3.1.1, starting at **step 17**). Immunoprecipitation is done with 1 mg of total microsomal proteins (*see* **Note 6**).

3.2 Early Readouts: ROS Burst, MAPK Activation, and Marker Gene Expression

3.2.1 ROS Burst

1. Grow *Arabidopsis* plants in soil under short-day conditions for 3–4 weeks.
2. Take leaf discs from 3- to 4-week-old *Arabidopsis* plants with the cork borer.
3. Take one leaf disc from each of the two youngest fully expanded leaves. Use at least 24 leaf discs per condition.
4. Place each leaf disc in the wells of a white 96-well plate containing 150 μL of water or the appropriate ligand treatment (*see* Fig. 1 for dosage) (*see* **Note 7**).
5. Keep the plate overnight at room temperature (*see* **Note 8**).
6. The following day, remove the water with a multichannel pipette without damaging the disks.
7. Add 100 μL of ROS detection solution and immediately measure the light emission with an appropriate luminometer (*see* Subheading 2.2, **step 2**).

3.2.2 MAPK Activation

1. Put sterile *Arabidopsis* seeds on plates containing the growth medium.
2. Stratify the seeds by keeping the plates in the dark at 4 °C for at least 2 days.
3. Transfer to long-day conditions and let seedlings grow for 5 days.
4. Transfer the seedlings to liquid 1/2MS medium. Single Eppendorf tubes and multi-well plates can be used; 24-well plates allow growing at least two seedlings per well. Use 6 seedlings per treatment.
5. Let the plants grow for another 7 days under long-day conditions.
6. The day before treatment and sampling, replace the liquid 1/2MS medium with 500 μL of fresh liquid 1/2MS medium.
7. Add 500 μL of flg22 and/or BL solution. Use respective solvents as negative control; wait for 5, 15, and 30 min. Blot dry and freeze in 1.5-mL Eppendorf tubes in liquid nitrogen. Strong MAPK activation usually peaks at 15 min.
8. Grind the seedlings in the 1.5-mL Eppendorf tubes with a micropestle. Do not let the samples thaw.

9. Add 100 μL of Lacus buffer to each tube. Mix thoroughly and keep the tubes on ice.
10. Centrifuge at more than 12,000 × *g* at 4 °C for 15 min.
11. Transfer the supernatant to a new tube and repeat the centrifugation step.
12. Quantify the protein concentration using Bradford and prepare aliquots with identical protein contents.
13. Add 5× protein loading buffer, boil for 10 min, and load samples on protein gels. Proceed to immunoblot blot according to the manufacturer's protocol for protein gel and antibody (*see* **Note 9**).

3.2.3 Target Gene Expression

1. Put sterile *Arabidopsis* seeds on plates containing the growth medium.
2. Stratify the seeds by keeping the plates in the dark at 4 °C for at least 2 days.
3. Transfer to light and let seedlings grow for 14 days under long-day conditions.
4. Harvest the desired amount of seedlings and infiltrate with a solution of flg22 and/or BL by vacuum infiltration for 5 min. Use the respective solvents as negative control. Make sure all seedlings are immersed in the treatment solutions.
5. Wait for the desired time. A typical time course ranges from 0 to 180 min. Typical up- or down-regulation of selected marker genes peaks between 60 and 120 min upon treatment.
6. Dry seedlings carefully and proceed to next step.
7. Extract total RNA by using favorite kit or protocol (*see* Subheading 2.2, **step 6**).
8. Perform cDNA first-strand synthesis using favorite kit or protocol.
9. Amplify DNA with preferred instrument and "target gene" primer pairs (*see* Subheading 2.2, **steps 8** and **9**).

3.3 Late Readouts: Callose Deposition and Seedling Growth Inhibition (SGI)

3.3.1 Callose Deposition

1. Put sterile *Arabidopsis* seeds on plates containing the growth medium.
2. Stratify the seeds by keeping the plates in the dark at 4 °C for at least 2 days.
3. Transfer seedlings under short-day condition and let seedlings grow for 14 days or until the first true leaves are fully expanded.
4. Submerge individual seedlings in 0.5-mL Eppendorf tubes containing a solution of flg22 at a concentration ranging between 1 and 10 μM.

5. Apply vacuum for 15 min.
6. Keep the seedlings in the elicitor solution (flg22) for another 16 h under short-day condition.
7. Transfer seedlings in 6-well plates containing 5 mL of fixing solution.
8. Remove fixing solution by pipetting. Do not damage the seedlings.
9. Add 70 % (v/v) ethanol and incubate for 2 h. Remove ethanol solution.
10. Add 50 % (v/v) ethanol and incubate for 2 h. Remove ethanol solution.
11. Add water and incubate overnight. Remove water solution.
12. Incubate seedlings in staining solution for several hours.
13. Mount individual true leaves on slides with 50 % (v/v) glycerol. At least 20 independent true leaves should be considered per treatment.
14. Observe callose immediately with a microscope under UV (excitation, 390 nm; emission, 460 nm).
15. Quantify callose deposition by counting callose dots.

3.3.2 Seedling Growth Inhibition

1. Put sterile *Arabidopsis* seeds on plates containing the growth medium.
2. Stratify the seeds by keeping the plates in the dark at 4 °C for at least 2 days.
3. Transfer seedlings under long-day light condition and let seedlings grow for 5 days.
4. Transfer the seedlings to a fresh batch of liquid MS medium containing the desired treatments (flg22 peptide and/or BL). Use respective solvents as negative control. Single Eppendorf tubes or multi-well plates can be used; 48-well plates accommodate one seedling per well. At least 16 seedlings per condition are enough. However, we recommend using 3-fold more seedlings per measurement. Concentrations of flg22 ranging from 10 to 100 nM usually provide a good enough resolution. When used in conjunction, BL is usually kept at 1000 nM.
5. Optional: At this point, a 5-min vacuum infiltration could be done to obtain more homogenous results.
6. Transfer back to long-day light condition and allow to grow for another 10 days.
7. Carefully remove the liquid medium by blotting dry.
8. Measure fresh weight of each seedling with a precision balance.

4 Notes

1. In our assays, we use *Arabidopsis* material from plants grown under both long-day (16 h light–8 h dark) and short-day (8 h light–16 h dark) conditions at 21 °C. Variations in light regime can be made, but these changes might alter the assay readouts positively or negatively. If you chose one specific light condition, use it consistently.
2. BL is a poorly soluble molecule and, as such, is usually dissolved in ethanol or DMSO. Control the solvent effects appropriately during the assays.
3. 24-epibrassinolide (EpiBL) can substitute BL. However, because EpiBL is less potent than BL, take this into account when designing cotreatment experiments with flg22.
4. Coimmunoprecipitation assays can be done with native anti-BRI1, anti-FLS2, and anti-BAK1 antibodies. They are usually obtained upon request from the laboratories that produced them. Some of them are commercially available (www. agrisera. com). Otherwise, stable transgenic lines with tagged variants of the proteins can be used, although in this case the functionality of the tagged version needs to be considered.
5. Western blot analysis of immunoprecipitates can be performed with various antibodies. The phosphorylation status of BAK1, FLS2, and BRI1 can be monitored with anti-phosphothreonine or anti-phosphotyrosine antibodies from Novagen.
6. This protocol also works for coimmunoprecipitations of FLS2-BAK1 complexes.
7. Flg22 can be substituted with other PAMPs. However, molarity recalibration assays need to be made before proceeding with these assays. ROS burst offers a nice quick way to calibrate the assays with another PAMP. The use of elf18 (active epitope of bacterial EF-Tu) in the same concentration range as that of flg22 will lead to similar results.
8. WARNING: The incubation time in water has been shown to influence the amplitude of the ROS bursts. In some cases, ROS burst assays carried out with or without water incubation can lead to opposite outputs. For further details, please see [21].
9. The protocol presented here is highly sensitive to the buffer, treatment conditions, and recommended usage of the antibody. As such, any deviations from the procedure will probably result in failure.

Acknowledgments

This work is funded by the Shanghai Center for Plant Stress Biology of the Chinese Academy of Sciences (to R.L.-D.) and is supported by funds from the Austrian Academy of Science through the Gregor Mendel Institute (to Y.B.).

References

1. Lozano-Durán R, Zipfel C (2015) Trade-off between growth and immunity: role of brassinosteroids. Trends Plant Sci 20:12–19
2. Belkhadir Y, Yang L, Hetzel J et al (2014) The growth–defense pivot: crisis management in plants mediated by LRR-RK surface receptors. Trends Biochem Sci 39:447–456
3. Pieterse CMJ, Van der Does D, Zamioudis C et al (2012) Hormonal modulation of plant immunity. Annu Rev Cell Dev Biol 28: 489–521
4. Belkhadir Y, Jaillais Y (2015) The molecular circuitry of brassinosteroid signaling. New Phytol 206:522–540
5. Zipfel C (2014) Plant pattern-recognition receptors. Trends Immunol 35:345–351
6. Macho AP, Zipfel C (2014) Plant PRRs and the activation of innate immune signaling. Mol Cell 54:263–272
7. Yang H, Yuan S, Luo Y et al (2013) Diverse responses are involved in the defence of *Arabidopsis thaliana* against *Turnip crinkle virus*. Z Naturforsch C 68:148–154
8. Albrecht C, Boutrot F, Segonzac C et al (2012) Brassinosteroids inhibit pathogen-associated molecular pattern--triggered immune signaling independent of the receptor kinase BAK1. Proc Natl Acad Sci U S A 109: 303–308
9. Belkhadir Y, Jaillais Y, Epple P et al (2012) Brassinosteroids modulate the efficiency of plant immune responses to microbe-associated molecular patterns. Proc Natl Acad Sci U S A 109:297–302
10. Jaillais Y, Belkhadir Y, Balsemão-Pires E et al (2011) Extracellular leucine-rich repeats as a platform for receptor/coreceptor complex formation. Proc Natl Acad Sci U S A 108: 8503–8507
11. Lozano-Durán R, Macho AP, Boutrot F et al (2013) The transcriptional regulator BZR1 mediates trade-off between plant innate immunity and growth. Elife 2:e00983
12. Jiménez-Gongora T, Kim S-K, Lozano-Durán R et al (2015) Flg22-triggered immunity negatively regulates key BR biosynthetic genes. Front Plant Sci 6:981
13. Lin W, Lu D, Gao X et al (2013) Inverse modulation of plant immune and brassinosteroid signaling pathways by the receptor-like cytoplasmic kinase BIK1. Proc Natl Acad Sci U S A 110:12114–12119
14. Shi H, Shen Q, Qi Y, Yan H, Nie H, Chen Y, Zhao T, Katagiri F, Tang D (2013) BR-SIGNALING KINASE1 physically associates with FLAGELLIN SENSING2 and regulates plant innate immunity in *Arabidopsis*. Plant Cell 25:1143–1157
15. Oñate-Sánchez L, Vicente-Carbajosa J (2008) DNA-free RNA isolation protocols for *Arabidopsis thaliana*, including seeds and siliques. BMC Res Notes 1:93
16. Gampala SS, Kim T-W, He J-X et al (2007) An essential role for 14-3-3 proteins in brassinosteroid signal transduction in *Arabidopsis*. Dev Cell 13:177–189
17. Malinovsky FG, Batoux M, Schwessinger B et al (2014) Antagonistic regulation of growth and immunity by the Arabidopsis basic helix-loop-helix transcription factor homolog of brassinosteroid enhanced expression2 interacting with increased leaf inclination1 binding bHLH1. Plant Physiol 164:1443–1455
18. Nole-Wilson S, Rueschhoff EE, Bhatti H, Franks RG (2010) Synergistic disruptions in *seuss cyp85A2* double mutants reveal a role for brassinolide synthesis during gynoecium and ovule development. BMC Plant Biol 10:198
19. Hartwig T, Corvalan C, Best NB et al (2012) Propiconazole is a specific and accessible brassinosteroid (BR) biosynthesis inhibitor for *Arabidopsis* and maize. PLoS One 7:e36625
20. Nemhauser JL, Hong F, Chory J (2006) Different plant hormones regulate similar processes through largely nonoverlapping transcriptional responses. Cell 126:467–475
21. Tateda C, Zhang Z, Shrestha J et al (2014) Salicylic acid regulates *Arabidopsis* microbial pattern receptor kinase levels and signaling. Plant Cell 26:4171–4187

Chapter 7

Identification of Brassinosteroid Target Genes by Chromatin Immunoprecipitation Followed by High-Throughput Sequencing (ChIP-seq) and RNA-Sequencing

Trevor Nolan, Sanzhen Liu, Hongqing Guo, Lei Li, Patrick Schnable, and Yanhai Yin

Abstract

Brassinosteroids (BRs) play important roles in many growth and developmental processes. BRs signal to regulate BR-INSENSITIVE1-ETHYL METHANESULFONATE-SUPPRESSOR1 (BES1) and BRASSINAZOLE-RESISTANT1 (BZR1) transcription factors (TFs), which, in turn, regulate several hundreds of transcription factors (termed BES1/BZR1-targeted TFs or BTFs) and thousands of genes to mediate various BR responses. Chromatin Immunoprecipitation followed by high-throughput sequencing (ChIP-seq) with BES1/BZR1 and BTFs is an important approach to identify BR target genes. In combination with RNA-sequencing experiments, these genomic methods have become powerful tools to detect BR target genes and reveal transcriptional networks underlying BR-regulated processes.

Key words Transcription factor, Target genes, ChIP-seq, RNA-seq, Gene expression

1 Introduction

Brassinosteroids (BRs) are an important group of plant steroid hormones that regulate numerous processes, including growth and development as well as both biotic and abiotic responses [1, 2]. BRs are perceived by the plasma membrane localized receptor BRASSINOSTEROID INSENSITIVE1 (BRI1) and coreceptor BRI1-ASSOCIATED KINASE1 (BAK1) and several other signaling components; ultimately, BR signaling activates BR-INSENSITIVE1-ETHYL METHANESULFONATE-SUPPRESSOR1/BRASSINAZOLE-RESISTANT1 (BES1/BZR1) family transcription factors (TFs) [3–6]. BES1 and BZR1 bind to the promoters of over 6600 target genes [7, 8], including several hundred BES1/BZR1-targeted TFs (BTFs), directing a transcriptional network that controls the expression of approximately 5000 BR-regulated genes under various growth stages and

Eugenia Russinova and Ana I. Caño-Delgado (eds.), *Brassinosteroids: Methods and Protocols*, Methods in Molecular Biology, vol. 1564, DOI 10.1007/978-1-4939-6813-8_7, © Springer Science+Business Media LLC 2017

conditions. Understanding how BES1 and BTFs control BR-regulated gene expression requires information regarding the genes bound and regulated by specific TFs in the signaling network.

RNA sequencing (RNA-seq) and chromatin immunoprecipitation followed by high-throughput sequencing (ChIP-seq) are now established as powerful techniques to determine genes that are transcriptionally regulated and directly bound by a particular TF, respectively [9, 10]. To study BR-regulated genes by RNA-seq, plants are grown and treated with or without the most active BR, brassinolide (BL). RNA is extracted and submitted for library preparation and sequencing by means of the Illumina platform. Genes regulated by a TF can be determined by performing RNA-seq on loss-of-function mutants for a TF (or its family members) and/or gain-of-function overexpression lines. Similarly, ChIP-seq with antibodies generated against a TF or with tagged TF lines (e.g., MYC, FLAG, or green fluorescent protein [GFP]) can reveal the direct targets of that TF. ChIP is accomplished by cross-linking DNA–protein complexes in vivo with formaldehyde, performing nuclear isolation followed by sonication to sheer the chromatin, and, subsequently, immunoprecipitating the DNA–TF complex with specific antibodies to the tag or TF. The DNA can then be reverse cross-linked, purified, and validated by ChIP-quantitative polymerase chain reaction (qPCR) of known target genes before library preparation and next-generation sequencing. The high-throughput sequencing data can be analyzed by bioinformatics and statistic tools. In this chapter, the RNA-seq and ChIP-seq protocols to identify BR target genes are described and an overview of the methods for data analyses is presented as well.

In combination, RNA-seq and ChIP-seq can be used to determine the genes a particular TF binds to and regulates. In the case of BTFs, these studies provide additional insight into how BR signaling is carried out and fine-tuned, but this methodology is generally applicable to any signaling pathway or process of interest. Genome-wide ChIP experiments in combination with RNA-seq are used to identify the targets for both BES1 and BZR1 [7, 8, 11]. When combined, approximately 27 % or 1765/6629 genes bound by BES1 and/or BZR1 are also regulated by BRs (Fig. 1).

The genes that are both bound and regulated by a TF have been considered the gold standard in ChIP studies, but recent work has demonstrated a hit-and-run model that occurs for at least some TFs [12, 13]. In this situation, genes are transiently bound by a TF and are activated, with dissociation of the TF, meaning that genes that are regulated, but not stably bound, could still be true targets of a TF. In any case, combining ChIP-seq and RNA-seq can help determine both the class of target genes that are stably bound and regulated by a TF and the genes that are regulated but

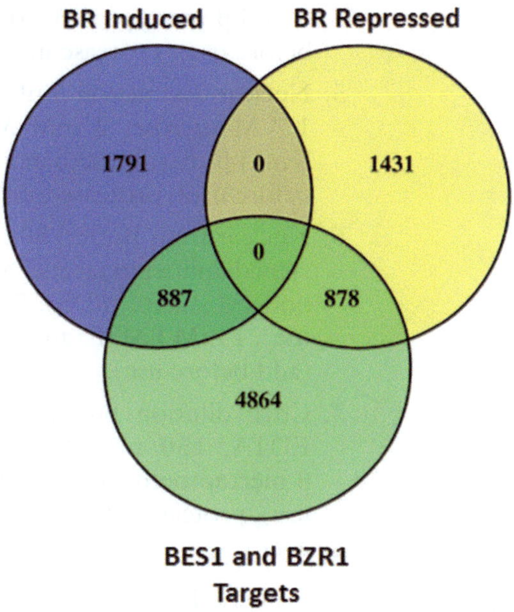

Fig. 1 Comparison of BR-regulated genes from RNA-seq and BES1 and BZR1 target genes identified by ChIP [7, 8, 11].

not bound by a TF. The latter either represents indirect regulation by other TFs or transiently bound targets, which can be distinguished by additional experiments. The protocols presented are based on published work and our own experimentation [14–16].

2 Materials

2.1 Plant Materials

1. *Arabidopsis thaliana* (L.) Heynh. seeds, including wild type (Columbia-0 accession), overexpression lines, and mutant lines for the TF of interest.
2. Plant growth trays and pots.
3. Soil.
4. Plant growth chamber with 16 h light–8 h dark regime and 20–22 °C temperature.

2.2 ChIP Solutions, Kits, and Reagents

1. 1 % (v/v) formaldehyde.
2. 2 M glycine.
3. Nuclear extraction buffer A: 10 mM Tris–HCl (pH 8.0), 0.4 M sucrose, 5 mM β-mercaptoethanol (add before use), 1 mM phenylmethylsulfonylfluoride (PMSF) (add before use), protease inhibitor cocktail (add before use).
4. Nuclear extraction buffer B: 10 mM Tris–HCl (pH 8.0), 0.25 M sucrose, 10 mM $MgCl_2$, 1 % (v/v) Triton X-100,

5 mM β-mercaptoethanol (add before use), 1 mM PMSF (add before use), protease inhibitor cocktail (add before use).

5. Nuclear Extraction Buffer C: 10 mM Tris–HCl (pH 8.0), 1.7 M sucrose, 2 mM $MgCl_2$, 0.15 % (v/v) Triton X-100, 5 mM β-mercaptoethanol (add before use), 1 mM PMSF (add before use), protease inhibitor cocktail (add before use).
6. Nuclear Lysis Buffer: 50 mM Tris–HCl (pH 8.0), 10 mM ethylenediaminetetraacetic acid (EDTA), 0.5 % (v/v) sodium dodecyl sulfate (SDS), 5 mM β-mercaptoethanol (add before use), 1 mM PMSF (add before use), protease inhibitor cocktail (add before use).
7. ChIP dilution buffer: 20 mM Tris–HCl (pH 8.0), 2 mM EDTA, 150 mM NaCl, 1 % (v/v) Triton X-100, 5 mM β-mercaptoethanol (add before use), 1 mM PMSF (add before use), protease inhibitor cocktail (add before use).
8. 5 M NaCl.
9. Protease K.
10. Antibodies: control immunoglobulin G (IgG) and antibody specifically recognizing the TF of interest or a tag (e.g., MYC, FLAG, or GFP) fused to the TF.
11. Dynabeads for protein A or protein G.
12. Low-salt wash buffer: 20 mM Tris–HCl (pH 8.0), 2 mM EDTA, 150 mM NaCl, 1 % (v/v) Triton X-100, 0.005 % (w/v) SDS.
13. High-salt wash buffer: 20 mM Tris–HCl (pH 8.0), 2 mM EDTA, 500 mM NaCl, 1 % (v/v) Triton X-100, 0.005 % (w/v) SDS.
14. TE buffer: 10 mM Tris–HCl (pH 8.0), 1 mM EDTA.
15. Chelating resin, such as Chelex 100 Resin (Bio-Rad).
16. ChIP elution buffer (freshly prepared): 0.1 M $NaHCO_3$, 1 % (w/v) SDS.
17. Glycogen.
18. SYBR Green PCR Master Mix.
19. ChIP DNA Clean & Concentrator Kit.
20. Illumina TrueSeq ChIP Sample Preparation Kit or Bio Scientific NEXTflex ChIP-seq Kit.

2.3 RNA-seq Solutions, Kits, and Reagents

1. 1 mM brassinolide: prepare stock in dimethylsulfoxide (DMSO) and store at −20 °C.
2. DMSO.
3. TRIzol.
4. Chloroform.

5. Isopropanol.
6. 75 % (v/v) ethanol made with RNase-free water.
7. RNase-free water.
8. RNase-Free DNase Set (Qiagen), including buffer RDD and DNase I stock solution used in DNase treatment.
9. Plant RNeasy Mini kit (Qiagen), including spin columns, 2-mL collection tubes, and buffers RLT and RPE used in RNA cleanup.

2.4 Equipment and Other Materials

1. 50-mL conical centrifuge tubes.
2. Miracloth.
3. Benchtop centrifuge equipped with a rotor capable of spinning 50-mL tubes.
4. Water bath sonicator.
5. Probe sonicator.
6. 1.5-mL low retention siliconized microcentrifuge tubes.
7. Magnetic stand.
8. Microcentrifuge.
9. Mx4000 Real-time PCR machine.
10. Mortar and pestle.
11. NanoDrop ND-1000 UV-Vis Spectrophotometer (Thermo Fisher Scientific) or 2100 Bioanalyzer (Agilent).
12. Vacuum pump.
13. Tube rotator.

3 Methods

3.1 ChIP-sequencing (ChIP-seq)

3.1.1 Preparation and Cross-Linking of Plant Materials

1. Grow healthy 4- to 5-week-old plants under long-day conditions (16 h light–8 h dark) at 20–22 °C. Include wild-type control plants as well as loss-of-function (T-DNA insertion or RNA interference [RNAi] mutant) and/or overexpression lines (*see* **Note 1**). Plants should be randomly distributed in growth flats to avoid differences due to environment.
2. Collect 4 g of plant leaf tissue for each genotype in a 50-mL tube. Rinse the plants with distilled water to remove residual soil. Add 37 mL of 1 % (v/v) formaldehyde to submerge the plant tissue. Apply vacuum for 15 min (release/reapply vacuum once at 7.5 min to help infiltration of fixing solution to plant tissue). For large amount of plants, *see* **Note 2**.
3. Add 2.5 mL of 2 M glycine to stop cross-linking and apply vacuum for 10 min.

4. Wash the fixed tissue with distilled water (three times). Dry the tissue gently with paper towels. Freeze the tissue in liquid nitrogen wrapped in aluminum foil. The tissue can be processed immediately or stored at −80 °C.

3.1.2 Chromatin Isolation and Sonication

1. Grind the tissue to a fine powder (*see* **Note 3**) with a medium-size mortar and pestle. Add 40 mL of freshly prepared nuclear extraction buffer A to the powder. Mix with a spatula and vortex at high-speed for 10 s.
2. Filter the mixture through two layers of Miracloth into a new 50-mL tube (*see* **Note 4**).
3. Centrifuge at 2000 × *g* for 20 min in a benchtop centrifuge.
4. Remove the supernatant and resuspend the pellet with wide-mouth pipette tips (can cut the regular tips with scissors) in 1.2 mL freshly prepared Nuclear extraction buffer B. Centrifuge in a microfuge for 10 min at 12,000 × *g* at 4 °C.
5. Remove the supernatant and resuspend the pellet in 0.4 mL freshly prepared nuclear extraction buffer C. Centrifuge in a microfuge for 60 min at 12,000 × *g* at 4 °C.
6. Remove the supernatant and resuspend the nuclear pellet with 0.2 mL freshly prepared nuclear lysis buffer by pipetting. Add 0.4 mL freshly prepared ChIP dilution buffer. Mix well and keep on ice.
7. Sonicate the chromatin with a water bath sonicator in icy water for 30 cycles with 30 s-on and 30 s-off cycle. Replace the icy water after 15 min to prevent overheating of the chromatin samples.
8. Centrifuge at maximum speed for 10 min. Transfer 0.5 mL supernatant to a new tube. The chromatin can be stored at −80 °C at this stage.
9. Take 10 μL sonicated chromatin. Add 15 μL of ChIP dilution buffer and 1 μL 5 M NaCl, treated with 1 μL protease K at 50 °C for 30 min. Reverse cross-linking at 65 °C for 5 h and check the DNA size with 2 % (w/v) agarose gel (Fig. 2). For typical sonication results and use of alternative probe sonication, *see* **Note 5**.

3.1.3 Chromatin Immunoprecipitation and Washing

1. Dilute 0.5 mL chromatin solution ten times by adding 4.5 mL of ChIP dilution buffer. Use 0.8 mL to 1 mL for each immunoprecipitation reaction. From this step on, use 1.5-mL siliconized tubes.
2. Add 2–5 μg of IgG or specific antibody to each tube containing the diluted chromatin solution. Rotate at 4 °C overnight. For three genotypes, six samples need to be processed for each biological repeat.

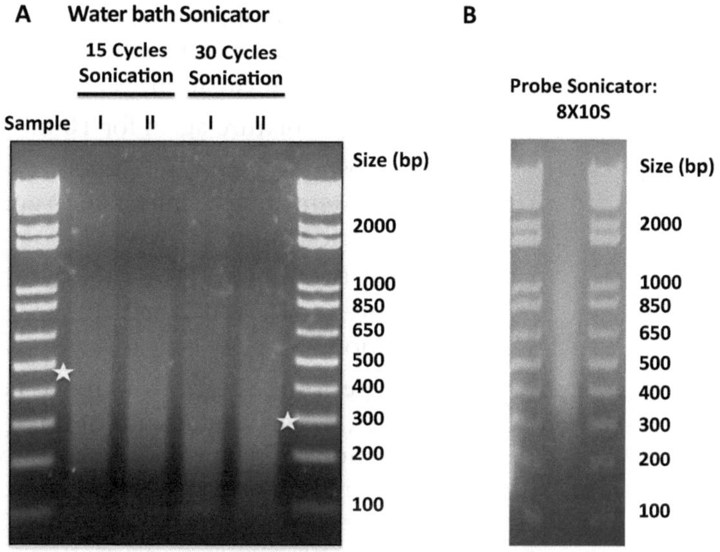

Fig. 2 Comparison of sonication methods for ChIP. (**a**) Sonication of chromatin for 15 or 30 cycles in a water bath sonicator. After 30 cycles, the average size of the chromatin fragments shifts to a decreased size (indicated with stars). (**b**) Sonication with a probe sonicator (power setting 2, 8 × 10 s, 1 min cooling).

3. For six samples, dilute 0.2 mL protein A Dynabeads (*see* **Note 6**) with 1 mL of ChIP dilution buffer in the 1.5-mL tubes by placing the bead-containing tubes in a magnetic stand for 20 s. Carefully remove the supernatant. Add again 1 mL of ChIP dilution buffer. Mix by inverting and precipitate the beads again. Add ChIP dilution buffer to 0.2 mL. The Dynabeads should be scaled up, when handling several biological repeats at the same time.
4. Add 30 μL of Dynabeads to each ChIP sample. Rotate at 4 °C in a tube rotator for 4 h.
5. Precipitate the Dynabeads with the ChIP complex as described above. Remove the supernatant and wash the beads with 1 mL of cold low-salt washing buffer.
6. Wash the Dynabeads with 1 mL of cold high-salt washing buffer (*see* **Note 7**).
7. Wash the beads twice with 1 mL of cold TE buffer and carefully remove any residual TE.

3.1.4 Elution of DNA and qPCR Analysis (See Note 8)

1. Add 0.1 mL 10 % (w/v) freshly prepared Chelex resin (1 g in 10 mL sterile water) to the Dynabeads (Subheading 3.1.3, **step 7**). Mix by vortexing.
2. Boil the mixture for 10 min to reverse cross-link the DNA–protein complexes. Mix the beads two to three times during the boiling process. Cool to room temperature.

3. Add 1 μL of protease K (10 μg/μL) to each tube and incubate at 50 °C in a water bath for 30 min. Mix three to four times during the incubation.
4. Boil the mixture again for 10 min to inactivate protease K.
5. Centrifuge 5 min at 12,000 × *g* at room temperature in a microcentrifuge. Collect the supernatant into a new tube as the first elution.
6. Add 0.1 mL of TE to the beads. Vortex and centrifuge as above. Collect 0.1 mL of supernatant and add to the first elution.
7. To perform qPCR with known target genes of a TF, for each gene, design a minimum of three pairs of primers, at different positions of the promoter region, with two pairs near potential TF-binding sites and one 3–5 kb upstream of the transcription start site as a negative control.
8. For each sample, use 2–4 μL of the purified DNA in each PCR reaction with 2–3 PCR reactions (technical repeats) per sample.
9. Run the PCR reactions as follows: 1 cycle at 95 °C for 10 min to activate the Taq polymerase, 40 cycles at 95 °C for 30 s, at 55 °C for 1 min, and at 72 °C for 30 s to amplify the DNA and collect fluorescence data, finally 1 cycle at 95 °C for 1 min and 55 °C for 30 s to obtain a melting curve for quality control of the primer pairs and PCR reactions.
10. Set the threshold to obtain a threshold cycle (Ct) for each reaction. The threshold is where significant and specific amplification occurs. Use the housekeeping gene ubiquitin 5 (UBQ5) as a reference gene to normalize the abundance of target genes. The normalized target gene abundance is calculated by delta threshold cycle (ΔCt).

    $\Delta Ct1 = Ct$ (target gene treated) $- Ct$ (UBQ5 treated).

    $\Delta Ct2 = Ct$ (target gene control) $- Ct$ (UBQ5 control).

    $\Delta\Delta Ct = \Delta Ct1 - \Delta Ct2$.

    the normalized abundance of the target gene $= 2^{\Delta\Delta Ct}$.
11. Compare the difference in enrichment of the promoter fragments by comparing $2^{\Delta\Delta Ct}$: the higher the $2^{\Delta\Delta Ct}$, the more enrichment of the specific promoter fragment.

3.1.5 Elution of DNA for ChIP-seq (See Note 8)

1. Elute DNA from Dynabeads (Subheading 3.1.3, **step** 7) with 0.25 mL of elution buffer at 65 °C. Repeat the elution by adding a second aliquot of 0.25 mL of elution buffer.
2. Combine the two elutions. Add 20 μL 5 M of NaCl and reverse cross-linking by incubating at 65 °C overnight.

3. Add 1 μL of protease K and incubate at 50 °C for 1 h.
4. Purify DNA with the ChIP DNA Clean and Concentrator Kit. For an alternative purification method, *see* **Note 9**.
5. Amplify the ChIP products with the Illumina TrueSeq ChIP Sample Preparation Kit or Bio Scientific NEXTflex ChIP-seq Kit according to the manufacturer's protocols. The sequencing is done with an Illumina Hi-seq 2500 with 50-bp reads.

3.2 RNA Sequencing (RNA-seq)

3.2.1 Plant Growth and BL Treatment

1. Grow plants as described for ChIP-seq (*see* Subheading 3.1.1).
2. Treat plants with 1 μM of BL prepared in water or with the same amount of DMSO (mock treatment) by gently spraying (approximately 50 mL for one tray of plants).
3. After 2.5 h, flash-freeze tissue in liquid nitrogen, pooling whole rosettes of at least three to four randomly chosen plants per sample. Harvest enough samples for at least three biological replicates for each genotype and treatment.

3.2.2 RNA Extraction

1. Grind tissues to a fine powder using liquid nitrogen, a mortar, and pestle. Take care that the tissue stays frozen during the grinding process by adding more liquid nitrogen as needed.
2. Using a prechilled spatula (in liquid nitrogen), transfer 100 mg of the tissue powder to a precooled 1.5-mL tube. Extra tissue powder can be stored separately at −80 °C for backup or future processing (*see* **Note 10**).
3. Add 1 mL of TRIzol reagent to the 1.5-mL tube containing the frozen tissue powder. Shake vigorously until the mixture is well resuspended (*see* **Note 11**).
4. Incubate the sample at room temperature for 5 min.
5. Add 0.2 mL of chloroform and shake the tube by hand for 15 s.
6. Incubate the tube for 2–3 min at room temperature, then centrifuge at 12,000 × *g* for 15 min at 4 °C in a microcentrifuge.
7. Remove the upper aqueous phase and place it in a new 1.5-mL tube, taking care not to disturb the interphase.
8. Add 0.5 mL of 100 % (v/v) isopropanol to the newly transferred aqueous phase and incubate at room temperature for 10 min.
9. Centrifuge at 12,000 × *g* for 10 min at 4 °C in a microcentrifuge.
10. Carefully remove the supernatant from the tube and wash the pellet by adding 1 mL of 75 % (v/v) ethanol.
11. Vortex the sample briefly and centrifuge 7500 × *g* for 5 min at 4 °C.

12. Remove the supernatant and allow the pellet to air dry for 5–10 min. Be careful not to overdry the pellet, because it makes it difficult to resuspend it completely.
13. Resuspend RNA in 87.5 μL of RNase-free water by heating for 10–15 min at 55 °C.

3.2.3 DNase Digestion

1. Perform DNAse digestion with RNase-Free DNase Set by adding 10 μL buffer RDD and 2.5 μL DNase I stock solution to each tube.
2. Incubate the tubes at room temperature for 10 min.

3.2.4 RNA Clean Up using RNeasy Mini Kit

1. Add 0.35 mL of buffer RLT to each tube. Mix well, add 0.25 mL 100 % (v/v) ethanol, and mix again.
2. Load the sample into RNeasy Mini spin column placed in a 2-mL collection tube. Centrifuge at approximately 10,000 × *g* for 15 s. Discard the flow through.
3. Add 0.5 mL of buffer RPE to the column. Centrifuge at approximately 8000 × *g* for 15 s. Discard the flow through.
4. Add another 0.5 mL of buffer RPE to the column. Centrifuge at approximately 8000 × *g* for 2 min. Discard the flow through.
5. Place the column in a new 2-mL collection tube and centrifuge at maximum speed for 1 min to remove any residual buffer RPE.
6. Transfer the column to a new 1.5-mL collection tube and elute the RNA by adding 30–50 μL of RNase-free water directly onto the spin column membrane. Centrifuge 1 min at approximately 8000 × *g*.
7. Assess the quality and concentration of the RNA with a bioanalyzer (NanoDrop or Agilent) (*see* **Note 12**).

3.2.5 Library Preparation

After RNA isolation and quality control, samples can be used for library preparation. The type of library preparation used will depend on the sequencing platform and conditions. Several protocols or kits are available. Alternatively, RNA can be submitted directly to a core facility or sequencing provider for library preparation and sequencing.

3.3 Bioinformatics Analyses of RNA-seq and ChIP-seq Data

RNA-seq and ChIP-seq analyses share similar procedures for data quality assessment, sequence trimming, alignment, and read counting. The sections overview the protocol for sequencing data for the RNA-seq and ChIP-seq experiments generated by the Illumina technology.

3.3.1 Data Quality Assessment

RNA-seq and ChIP-seq generate massive amounts of sequencing data that are generally delivered in FASTQ format (en.wikipedia.org/wiki/FASTQ_format). The quality of raw sequencing data can be assessed with FastQC (bioinformatics.babraham.ac.uk/projects/fastqc) that provides graphs and summary tables to overview data amount, quality, and adaptor contamination.

3.3.2 Quality and Adaptor Trimming

Raw data generated from a sequencing platform contain low-quality bases, typically enriched at the 3′-ends of each sequence, and adaptor sequences. The precise removal of low-quality sequences and adaptor contamination is important for downstream analyses. Many software tools have been developed for these purposes [17]. Trimmomatic [18] is a commonly used as trimming software that includes two modes for adaptor trimming: simple and palindrome mode. The former mode is universal for data from any sequencing platform, whereas the latter is designed for paired-end sequences. Palindrome mode has considerable advantages in sensitivity and specificity. After processing, the trimmed data are still in FASTQ format for subsequent procedures.

3.3.3 Alignment of Reads to a Reference Genome

For both RNA-seq and ChIP-seq, trimmed sequencing reads are typically aligned to a reference genome. For RNA-seq, reads are largely generated from mature mRNAs that are subjected to splicing. Therefore intron-tolerant aligners, such as GSNAP [19], STAR [20], and Tophat [21], should be used for alignments. For ChIP-seq, reads are from genomic DNAs. Either BWA [22] or Bowtie (or Bowtie2) [23, 24] can be applied. Both aligners are fairly fast and accurate. Owing to genome repetitiveness, reads can align to multiple genomic locations. Also some reads are aligned with too many mismatches, potential signaling misalignments. To reduce misalignments, postalignment filtering is used. A simple filtering is to set a minimal alignment quality score.

Sequence Alignment/Map format (SAM) is a generic alignment format employed in all next-generation sequencing programs to store alignments [25]. As SAM is a readable text file that is not storage efficient, it is usually converted to a BAM format that contains the same information, but is in a binary format, greatly enhancing storage efficiency. The conversion is implemented in SAMtools [25] that also provides an indexing function for BAM to improve the efficiency of further alignment access.

3.3.4 Alignment Visualization

Integrative Genomics Viewer (IGV) [26] is a flexible and user-friendly tool to enable visualization of diverse genomic data, including BAM alignment data. Multiple alignment data sets from different samples or different experiments (e.g., RNA-seq and ChIP-seq) can be displayed in the same panel. Other genomic data, such as genome annotation and variants, can be provided as well for visual exploration of the relationship among different genomic features.

3.3.5 Read Depth (Counts) at Genomic Regions

Read depth of genomic regions is determined based on the overlap between coordinates of reads mapped in the reference genome and genomic locations of target regions. Existing tools are available to compute read depth per genomic region. Two recommended software packages are HTSeq [27] and bedtools (bedtools.readthedocs.org). HTSeq is tailored for counting RNA-seq reads. It is flexible to count reads in defined genic regions, such as exons, genes, or coding regions, provided in GFF or GTF genome annotation files. It also offers options to deal with the cases in which gene models in a reference genome overlap. Bedtools is a powerful tool to cross-check multiple files in diverse data formats, such as BAM, BED, and GFF/GTF, enabling the counting of reads in defined genomic regions and calculating sequence coverage of genomic regions. Genomic regions can be genes that are used in RNA-seq or defined windows (or bins) used in ChIP-seq. Once count data are generated, normalization and statistical tests will be applied to identify differentiation between biological groups at genomic regions of biological interest.

3.3.6 Data Normalization and Statistical Tests for RNA-seq

The goal of RNA-seq is to identify genes that are differentially expressed (DE) among samples that differ in some known manner, genotype, tissue type, treatment, etc. To this end, read counts are compared for given genes across samples. Differences in sequencing depths, technical variation, and biological variation are all sources for variation observed in read counts per unit (for instance, a gene) among samples. For RNA-seq, a good experimental design is required to estimate different variation sources and declare the statistical significance of the biological factor of interest. To account for variation in sequencing depth, a normalization method, i.e., reads per kilobase per million reads (RPKM), has been widely used [28]. Similar to RPKM, fragments per kilobase of exon per million mapped reads (FPKM) was used later to avoid double counting of paired-end reads [29]. For RNA-seq, several statistical approaches for the identification of differential expression were developed in recent years, including DESeq [30, 31], edgeR [32], Cuffdiff [33], and baySeq [34]. Each of these software packages has its own algorithms for normalization and statistical modeling of gene expression. For all the software packages, a statistical test is performed for a gene or an isoform, resulting in an associated P value. When multiple-hypothesis tests are conducted, the same P value cutoff (for example, 0.05) for each test is not enough to guarantee the same level of false positives as that of a single-hypothesis test. To cope with this multiple testing problem, a false discovery rate (FDR) has been widely used [35]. As a further development the FDR, the Q value was introduced as an adjusted P value for the FDR determination [36]. The FDR estimates the expected fraction of incorrect rejections among all the rejections.

After a DE set of genes has been found, a test for enrichment of DE genes in certain families or functional groups can be performed. For example, the software GOseq [37] is used for a gene ontology (GO) enrichment analysis. *P* values for overrepresentation of DE genes for each GO category can be generated by comparison to a null distribution of each GO category from repeated samplings of the same number of DE genes. Although the multiple-testing issue exists, the FDR for this analysis should be applied with caution due to the complicated relationship among GO terms breaking the assumption of independence of each test in most FDR algorithms. During sampling, genes are weighted with their read counts from all samples to account for the potential bias in the differential power to detect DE among genes with different numbers of read counts. In addition, other pathway visualization tools, such as MapMan [38] or Kyoto Encyclopedia of Genes and Genomes (KEGG; genome.jp/kegg), can be used to better understand which pathways and functional modules are subjected to significant expression regulation.

3.3.7 Data Normalization and Statistical Tests for ChIP-seq

The goal of ChIP-seq is to identify the genomic regions to which an immunoprecipitated (IP) protein (such as a TF) is bound. The fixed chromatin is fragmented by sonication and immunoprecipitated. Small genomic DNA fragments are purified, amplified, and subjected to high-throughput sequencing. CHANCE [39] evaluates IP strength by separately plotting curves for both the IP sample(s) and the control(s), showing the empirical cumulative percentage of reads covered by the windows/bins that are ordered increasingly by their aligned reads. Potential quality problems, such as weak IP enrichment and insufficient sequencing depth, can be identified with outputs from CHANCE.

By analyzing read depth at windows/bins in both IP samples and controls, genomic regions with significant read enrichment are considered to be IP peaks that are potential protein-binding regions. Sharp and broad peaks are two major types of peak signals. Sharp peaks, often associated with protein–DNA binding domains, can be detected by various peak callers, such as MACS [40], CisGenome [41], SPP [42], CSAR [43], ZINBA [44], HPeak [45], and BayesPeak [46]. The detection of broad peaks resulting, for example, from signals associated with histone modifications, requires different algorithms, including SICER [47], CCAT [48], and RSEG [49], or sharp peak callers, such as MACS, adjusted with the appropriate parameters. Similar to RNA-seq, a FDR [35] is needed to account for multiple testing.

To produce reproducible results, biological replicates of a ChIP-seq experiment are needed. To measure the reproducibility between replicates, irreproducible discovery rate (IDR) analysis can be applied to peaks identified from two replicates [50, 51].

When biological replication is applied, RNA-seq software packages, such as DESeq and edgeR, can also be adapted to deal with biological replicates and identify statistically significant signals in ChIP samples versus their controls by means of the ChIP-seq read counts [52].

4 Notes

1. Most protocols use seedlings for ChIP experiments, but we found that BRs affect more genes at the adult stage, so mature plants are used in this protocol for both ChIP-seq and RNA-seq experiments. The use of appropriate controls in critical for ChIP experiments. When an antibody produced against a TF is used, the ideal control would be a loss-of-function mutant or an RNAi line for that TF. Alternatively, when a tagged line is used, it is important to include a wild-type control. Whenever possible, the tagged line should be in a mutant background to avoid competition with the endogenous TF. In this case, the mutant is used as a control.
2. Generally, 4 g of tissue is enough for most TFs studied. However for low-abundance or tissue-specific TFs, additional plant material should be used. A protocol for scaling up ChIP experiments while retaining low background has recently been described [14]. For large amount of plants (for instance, 20 g of tissue), a 500-mL flask with 185 mL 1 % (v/v) formaldehyde should be used and vacuum period should be increased to 20 min. To stop cross-linking in the next step, use 12.5 mL 2 M glycine.
3. We found that complete grinding is critical for good nuclear yield. For 4 g of tissue, we usually grind for approximately 15 min with plenty of liquid nitrogen, to be added three to four times in the process; for large amounts of tissue, such as 20 g, grinding should be done for 30 min.
4. Do not use pipette tips to facilitate flowing of the mixture through the filter. You can swirl the tube with Miracloth gently if needed, but do not filter more than 40 mL with one set of Miracloth.
5. Sonication is a critical step and needs to be optimized for each experiment and laboratory conditions. Parameters include the type of sonicator used, power setting, number of cycles, and SDS concentrations. Several settings should be tested and the results checked by DNA gel electrophoresis. We found that a water bath sonicator works better than a probe sonicator under our experimental conditions, but it is possible to obtain good results with both types of sonicators (Fig. 2).

6. Depending on the type of primary antibody, protein A or protein G beads may be used. New England Biolabs provides an excellent reference for choosing between these (https://www.neb.com/tools-and-resources/selection-charts/affinity-of-protein-ag-for-igg-types-from-different-species). In our experience, magnetic beads, such as Dynabeads, produce low background and a high signal-to-noise ratio, but traditional protein A agarose beads could be substituted and often result in a higher total signal.
7. For antibodies with low affinities, the salt concentration can be reduced from 500 mM to 250 mM or the high-salt wash could even be completely omitted.
8. This method uses Chelex resin and boiling to reverse cross-linking. The protocol is quicker (saves 1 day) than elution with the 1 % (w/v) SDS and 0.1 M $NaHCO_3$. However, we found that the boiling process denatures small DNA fragments that do not affect the qPCR analysis. Nevertheless, for ChIP-seq in which the ChIP product will be ligated to adaptors, the elution should be done with SDS and $NaHCO_3$ and reverse cross-linking at 65 °C.
9. The DNA can also be purified by phenol–chloroform extraction followed by ethanol precipitation. Glycogen should be added to facilitate DNA precipitation.
10. The plant material is sufficient for multiple RNA extractions and the ground tissue powder can be kept as backup at −80 °C and used later if samples do not meet the quality control standards.
11. All RNA extraction steps should be carried out with filter tips. Take care to avoid RNase contamination. We routinely clean all benches and supplies to be used for RNA extraction with the decontamination detergent RNase Away before starting the experiment.
12. Samples should have an optical density $(OD)_{260/280}$ and $OD_{260/230}$ greater than 1.8. The amount of RNA needed will depend on the library preparation method used, but samples should generally contain at least 400 ng RNA at a concentration of approximately 10 ng/μL.

Acknowledgments

We thank Iowa State University DNA Facility for allowing us to use the Diagenode Bioruptor and Mike Baker for information on ChIP-seq product amplification. The work is supported by grant from NSF (IOS-1257631) and by the Plant Science Institute at Iowa State University.

References

1. Clouse SD (2011) Brassinosteroid signal transduction: from receptor kinase activation to transcriptional networks regulating plant development. Plant Cell 23:1219–1230
2. Li J, Chory J (1999) Brassinosteroid actions in plants. J Exp Bot 50:275–282
3. Guo H, Li L, Aluru M et al (2013) Mechanisms and networks for brassinosteroid regulated gene expression. Curr Opin Plant Biol 16: 545–553
4. Zhu J-Y, Sae-Seaw J, Wang Z-Y (2013) Brassinosteroid signalling. Development 140: 1615–1620
5. Jiang J, Zhang C, Wang X (2013) Ligand perception, activation, and early signaling of plant steroid receptor brassinosteroid insensitive 1. J Integr Plant Biol 55:1198–1211
6. Vriet C, Russinova E, Reuzeau C (2013) From squalene to brassinolide: the steroid metabolic and signaling pathways across the plant kingdom. Mol Plant 6:1738–1757
7. Yu X, Li L, Zola J et al (2011) A brassinosteroid transcriptional network revealed by genome-wide identification of BESI target genes in *Arabidopsis thaliana.* Plant J 65: 634–646
8. Sun Y, Fan X-Y, Cao D-M et al (2010) Integration of brassinosteroid signal transduction with the transcription network for plant growth regulation in *Arabidopsis.* Dev Cell 19:765–777
9. Mundade R, Ozer HG, Wei H et al (2014) Role of ChIP-seq in the discovery of transcription factor binding sites, differential gene regulation mechanism, epigenetic marks and beyond. Cell Cycle 13:2847–2852
10. Soon WW, Hariharan M, Snyder MP (2013) High-throughput sequencing for biology and medicine. Mol Syst Biol 9:640
11. Wang X, Chen J, Xie Z et al (2014) Histone lysine methyltransferase SDG8 is involved in brassinosteroid-regulated gene expression in *Arabidopsis thaliana.* Mol Plant 7:1303–1315
12. Li Y, Mukherjee I, Thum KE et al (2015) The histone methyltransferase SDG8 mediates the epigenetic modification of light and carbon responsive genes in plants. Genome Biol 16:79
13. Para A, Li Y, Marshall-Colón A et al (2014) Hit-and-run transcriptional control by bZIP1 mediates rapid nutrient signaling in *Arabidopsis.* Proc Natl Acad Sci USA 111:10371–10376
14. Lau OS, Bergmann DC (2015) MOBE-ChIP: a large-scale chromatin immunoprecipitation assay for cell type-specific studies. Plant J 84: 443–450
15. Nelson JD, Denisenko O, Sova P et al (2006) Fast chromatin immunoprecipitation assay. Nucleic Acids Res 34:e2
16. Gendrel AV, Lippman Z, Yordan C et al (2002) Dependence of heterochromatic histone H3 methylation patterns on the *Arabidopsis* gene *DDM1*. Science 297:1871–1873
17. Chen C, Khaleel SS, Huang H et al (2014) Software for pre-processing Illumina next-generation sequencing short read sequences. Source Code Biol Med 9:8
18. Bolger AM, Lohse M, Usadel B (2014) Trimmomatic: a flexible trimmer for Illumina sequence data. Bioinformatics 30:2114–2120
19. Wu TD, Nacu S (2010) Fast and SNP-tolerant detection of complex variants and splicing in short reads. Bioinformatics 26:873–881
20. Dobin A, Davis CA, Schlesinger F et al (2013) STAR: ultrafast universal RNA-seq aligner. Bioinformatics 29:15–21
21. Trapnell C, Pachter L, Salzberg SL (2009) TopHat: discovering splice junctions with RNA-Seq. Bioinformatics 25:1105–1111
22. Li H, Durbin R (2009) Fast and accurate short read alignment with Burrows-Wheeler transform. Bioinformatics 25:1754–1760
23. Langmead B (2010) Aligning short sequencing reads with Bowtie. Curr Protoc Bioinformatics 32 Unit 11.7:1–14
24. Langmead B, Salzberg SL (2012) Fast gapped-read alignment with Bowtie 2. Nat Methods 9:357–359
25. Li H, Handsaker B, Wysoker A et al (2009) The sequence alignment/map format and SAMtools. Bioinformatics 25:2078–2079
26. Robinson JT, Thorvaldsdóttir H, Winckler W, Guttman M, Lander ES, Getz G, Mesirov JP (2011) Integrative genomics viewer. Nat Biotechnol 29:24–26
27. Anders S, Pyl PT, Huber W (2015) HTSeq--a Python framework to work with high-throughput sequencing data. Bioinformatics 31:166–169
28. Mortazavi A, Williams BA, McCue K et al (2008) Mapping and quantifying mammalian transcriptomes by RNA-seq. Nat Methods 5:621–628
29. Trapnell C, Williams BA, Pertea G et al (2010) Transcript assembly and quantification by RNA-Seq reveals unannotated transcripts and isoform switching during cell differentiation. Nat Biotechnol 28:511–515
30. Love MI, Huber W, Anders S (2014) Moderated estimation of fold change and dispersion

for RNA-seq data with DESeq2. Genome Biol 15:550
31. Anders S, Huber W (2010) Differential expression analysis for sequence count data. Genome Biol 11:R106
32. Robinson MD, McCarthy DJ, Smyth GK (2010) edgeR: a Bioconductor package for differential expression analysis of digital gene expression data. Bioinformatics 26:139–140
33. Trapnell C, Hendrickson DG, Sauvageau M et al (2013) Differential analysis of gene regulation at transcript resolution with RNA-seq. Nat Biotechnol 31:46–53
34. Hardcastle TJ, Kelly KA (2010) baySeq: empirical Bayesian methods for identifying differential expression in sequence count data. BMC Bioinformatics 11:422
35. Benjamini Y, Hochberg Y (1995) Controlling the false discovery rate: a practical and powerful approach to multiple testing. J R Stat Soc Ser B-Stat Methodol 57:289–300
36. Storey JD (2002) A direct approach to false discovery rates. J Roy Stat Soc Ser B-Stat Methodol 64:479–498
37. Young MD, Wakefield MJ, Smyth GK et al (2010) Gene ontology analysis for RNA-seq: accounting for selection bias. Genome Biol 11:R14
38. Thimm O, Bläsing O, Gibon Y et al (2004) MAPMAN: a user-driven tool to display genomics data sets onto diagrams of metabolic pathways and other biological processes. Plant J 37:914–939
39. Diaz A, Nellore A, Song JS (2012) CHANCE: comprehensive software for quality control and validation of ChIP-seq data. Genome Biol 13:R98
40. Zhang Y, Liu T, Meyer CA et al (2008) Model-based analysis of ChIP-Seq (MACS). Genome Biol 9:R137
41. Ji H, Jiang H, Ma W et al (2008) An integrated software system for analyzing ChIP-chip and ChIP-seq data. Nat Biotechnol 26:1293–1300
42. Kharchenko PV, Tolstorukov MY, Park PJ (2008) Design and analysis of ChIP-seq experiments for DNA-binding proteins. Nat Biotechnol 26:1351–1359
43. Muiño JM, Kaufmann K, van Ham RCHJ et al (2011) ChIP-seq analysis in R (CSAR): an R package for the statistical detection of protein-bound genomic regions. Plant Methods 7:11
44. Rashid NU, Giresi PG, Ibrahim JG et al (2011) ZINBA integrates local covariates with DNA-seq data to identify broad and narrow regions of enrichment, even within amplified genomic regions. Genome Biol 12:R67
45. Qin Z, Yu J, Shen J et al (2010) HPeak: an HMM-based algorithm for defining read-enriched regions in ChIP-Seq data. BMC Bioinformatics 11:369
46. Spyrou C, Stark R, Lynch AG et al (2009) BayesPeak: Bayesian analysis of ChIP-seq data. BMC Bioinformatics 10:299
47. Zang C, Schones DE, Zeng C et al (2009) A clustering approach for identification of enriched domains from histone modification ChIP-Seq data. Bioinformatics 25:1952–1958
48. Xu H, Handoko L, Wei X et al (2010) A signal-noise model for significance analysis of ChIP-seq with negative control. Bioinformatics 26:1199–1204
49. Song Q, Smith AD (2011) Identifying dispersed epigenomic domains from ChIP-Seq data. Bioinformatics 27:870–871
50. Bailey T, Krajewski P, Ladunga I et al (2013) Practical guidelines for the comprehensive analysis of ChIP-seq data. PLoS Comput Biol 9:e1003326. doi:10.1371/journal.pcbi.1003326
51. Li Q, Brown JB, Huang H et al (2011) Measuring reproducibility of high-throughput experiments. Ann Appl Stat 5:1752–1779
52. Furey TS (2012) ChIP-seq and Beyond: new and improved methodologies to detect and characterize protein-DNA interactions. Nat Rev Genet 13:840–852

Chapter 8

Quantitation of Cell Type-Specific Responses to Brassinosteroid by Deep Sequencing of Polysome-Associated Polyadenylated RNA

Kristina Vragović, Elizabeth Bartom, and Sigal Savaldi-Goldstein

Abstract

Hormonal signaling pathways control almost every aspect of plant physiology and development. Extensive analysis of hormonal signaling output, i.e., gene expression, has therefore been the focus of many studies. These analyses have been primarily conducted on total extracts derived from a mixture of tissues and cell types, consequentially limiting delineation of precise models. In this chapter, methods for tissue-specific functional genomics are overviewed, in which hormonal responses are analyzed at the transcriptional and the translational levels. Deep sequencing of polysome-associated polyadenylated RNA is employed for cell type-specific quantitation of translatome responses to brassinosteroids. Polysomes are purified by the previously established Translating Ribosome Affinity Purification (TRAP) method, in which the expression of a tagged ribosomal protein is targeted to the tissue of interest, allowing tissue-specific immunopurification of the polysome complexes. The methods presented assess establishment and selection of suitable transgenic lines. A protocol for hormonal treatment of the *Arabidopsis thaliana* root as a case study, TRAP and linear amplification of the purified polysome-associated polyadenylated RNA are described. Finally, a step-by-step presentation is included of the analysis of the RNA deep-sequencing data and Rscript for plotting hierarchically clustered heatmap of the expressed genes.

Key words Brassinosteroid, RNA-seq, Translatome, Root growth, Gene expression, Ribosome, Affinity purification, Hierarchical clustering, Functional genomics

1 Introduction

1.1 Identification of Hormonal Targets by Profiling Global Gene Expression in Specific Tissues at the Transcriptional Level

Hormonal target genes are largely identified by quantifying gene responses to temporal changes in hormone signal intensities, at different developmental stages, and under varying environmental conditions. A common strategy is therefore to compare gain- and loss-of-function mutants introduced into wild-type plants and their responses to plant hormone level perturbations, such as by application of the hormone over time. Numerous genome-wide approaches have been developed to profile specific plant tissues and

Eugenia Russinova and Ana I. Caño-Delgado (eds.), *Brassinosteroids: Methods and Protocols*, Methods in Molecular Biology, vol. 1564, DOI 10.1007/978-1-4939-6813-8_8,

cells (for a recent comprehensive review, see ref. [1]). Surprisingly, only a few studies exploited these tools when hormonal effects were investigated. These high-resolution techniques include Fluorescence Activated Cell Sorting (FACS) [2, 3], Isolation of Nuclei Tagged in specific Cell Type (INTACT) [4, 5], and Translating Ribosome Affinity Purification (TRAP) [6, 7]. In the first method, transgenic plants expressing a fluorescent protein in specific tissues are subjected to rapid enzymatic dissection of the cell wall and the resulting protoplasts are then sorted by FACS. This method was used to map auxin responses in different tissues of the *Arabidopsis thaliana* root [8] and to characterize the impact of auxin and ethylene on root hair formation [9]. FACS was also used to probe gene responses to brassinosteroid (BR) in root vasculature cells [10]. In INTACT, cell type-specific promoters drive the expression of a fluorescently labeled nuclear envelope-associated protein, the nuclear targeting fusion (NTF) protein. The biotin ligase BirA, broadly expressed by the NTF-expressing line, biotinylates NTF, allowing isolation of the NTF-tagged nuclei on streptavidin columns. Application of INTACT for hormonal studies has not been reported yet. Both FACS and INTACT are suitable for transcriptome analyses, but the transcript level does not necessarily correlate with the amount of protein synthesis, regulated at the initiation, elongation, and termination stages [11]. Hormonal signals; for example, have been shown to trigger gene-specific translational regulation, causing a discrepancy between transcript accumulation and its translational level [12, 13]. TRAP, the focus of this chapter, is an alternative method that provides a more accurate measure of translation (see below).

1.2 Identification of Hormonal Targets by Profiling Global Gene Expression at the Translational Level

The TRAP method enables the isolation of polysome-associated RNA (collectively referred to as the "translatome"), providing a good approximation of the protein abundance. The *Arabidopsis* ribosomal protein L18 (RPL18), fused to a dual short tag of 6× histidine and FLAG peptide, has been shown to efficiently incorporate into the cytosolic surface of the 60S subunit [6]. Once tagged, the entire polysomes and their associated RNAs can be immunopurified by commercially available anti-FLAG antibodies conjugated to agarose beads. FLAG-RPL18 expression under tissue-specific promoters allows recovery of different polyadenylated RNA species: messenger (m)RNA, alternatively spliced forms, and noncoding RNAs [14, 15], from the cell types of interest [7]. Thus, TRAP can be used not only as a tool for identification of hormonal control of translation, but also for recognition of cell-specific responses beyond protein expression [16]. Similarly, genome-wide ribosome footprinting (Ribo-seq) determines translational efficiency by comparing the number of ribosomes per associated mRNA relative to the total abundance of the transcript [12].

Ribo-seq, however, does not distinguish between translationally active and inactive ribosomes. The use of this method has uncovered gene-specific translational regulation by ethylene [17].

1.3 Profiling Tissue-Specific Gene Translation in Response to BR in Wild Type and BR Mutant Backgrounds by Means of the TRAP Method

BR signaling differentially controls organ growth and development, depending on the tissue perceiving the BR signal [18, 19]. RNA-seq of polysome-associated RNA from specific tissues of the *Arabidopsis* primary root expressing FLAG-RPL18 has provided a molecular underpinning for the discovery of opposing developmental programs triggered by BR activity in the outer and the inner root tissues [16, 19]. As the outer and inner tissues strikingly differ in their response to exogenous BR they are implicated in a distinct magnitude of responding genes, number of genes responding by repression and activation, expression patterns, and distribution in functional annotation categories. In addition, by crossing tissue-specific FLAG-RPL18 lines to genetic backgrounds with tissue-specific perturbations of the signaling pathways, autonomously and nonautonomously regulated genes have been distinguished [16].

Study of gene activation responses in specific cells and tissues requires prior knowledge of the promoter fragments that can specifically drive gene expression. For translatome analysis, the expected expression pattern should be maintained under different hormonal regimes and mutant backgrounds. Therefore, the use of binary vectors containing the 35S enhancer sequence (i.e., a vector with a 35S promoter-driven selection marker) should be avoided, because they trigger ectopic expression of the gene of interest [20]. Selection of a transgenic line expressing the FLAG-RPL18 is important for successful analysis of tissue-specific responses. To account for positional effects associated with the T-DNA integration into the genome, the new transgenic lines are first validated by quantitative reverse transcription polymerase chain reaction (qRT-PCR) of the isolated polysome-associated RNA. This comparison of the relative enrichment of known tissue-specific genes among independent lines (such as the endogenous gene corresponding to the promoter), allows selection of the most suitable line for further analysis. Once selected, the line is used as the parental genetic background for all genetic crosses. BRs are commonly applied at high, nonphysiological concentrations to ensure detection of genomic responses. Importantly, because transport mechanisms of BRs are unknown, penetration of the applied hormone to the plant inner cells should be verified, for example, by using plants expressing a fluorescent readout for the BR signaling pathways, such as BR INSENSITIVE1 KINASE INHIBOTOR1 and BR-INSENSITIVE1-ETHYL METHANESULFONATE-SUPPRESSOR1/BRASSINAZOLE-RESISTANT1 [16, 21, 22]. In addition to recommendations for appropriate selection of transgenic lines, this chapter presents a protocol for BR treatment of roots, root harvest, polysome immunopurification, extraction of polysome-associated RNA, and amplification of polyadenylated RNA (Fig. 1). Methods are also detailed for RNA-seq data analysis, including a link to an Rscript for a hierarchically clustered heatmap of genes.

2 Materials

2.1 Hormone Treatment

1. Seeds of selected transgenic lines expressing FLAG-tagged RPL18 (FLAG-RPL18) under tissue-specific promoters.
2. Desiccator for seed sterilization.
3. 100-mL beaker.
4. Sterile square petri plates with grid (100 × 100 × 15 mm) (*see* **Note 1**).
5. Tweezers, rinsed with 70 % (v/v) ethanol (EtOH).
6. 1.7-mL RNAse-free microfuge tubes.
7. 50-mL RNAse-free Falcon tubes. For chemical treatments exceeding 50 mL of final solution volume, use 250-mL Erlenmeyer flasks and 100- to 250-mL graduated glass cylinders, baked for 2 h at 200 °C.
8. Laboratory wipes, cut into long, 1-cm-wide strips, sufficient to fit a square petri plate. For treatments exceeding 8 h, wrap strips in aluminum foil and autoclave.
9. Pipetting tips, autoclaved.
10. Parafilm.
11. Half-strength Murashige–Skoog (½ MS) medium with 0.8 % (w/v) plant agar and 0.2 % (w/v) sucrose (4.3 g MS medium, 0.92 g 2-(*N*-morpholino)ethanesulfonic acid (MES) hydrate, and 4 g sucrose). Dissolve in 1.9 L double-distilled water. Bring volume to 2 L (sufficient for preparing 80 petri plates), and adjust pH to 5.8 with 1 M of NaOH. Distribute into 400-mL autoclavable bottles and add 3.2 g plant agar. Autoclave and store at 4–8 °C. Prior to use, bring the medium to a boil in a microwave oven and cool down to 40–50 °C. Pour 25 mL into each square petri plate under aseptic conditions and wait until solidified.
12. Liquid ½ MS medium with 0.2 % (w/v) sucrose. Store at 4–8 °C (*see* **Note 2**).
13. 1 mM brassinolide (BL, $C_{28}H_{48}O_6$), dissolved in 100 % (v/v) dimethyl sulfoxide (DMSO, C_2H_6OS), stored as 50-µL aliquots, and frozen at −20 °C. Prepare fresh on the day of the experiment (*see* **Note 3**).
14. 37 % (v/v) hydrochloric acid (HCl).
15. 100 % (v/v) dimethyl sulfoxide (DMSO).
16. 70 % (v/v) ethanol.

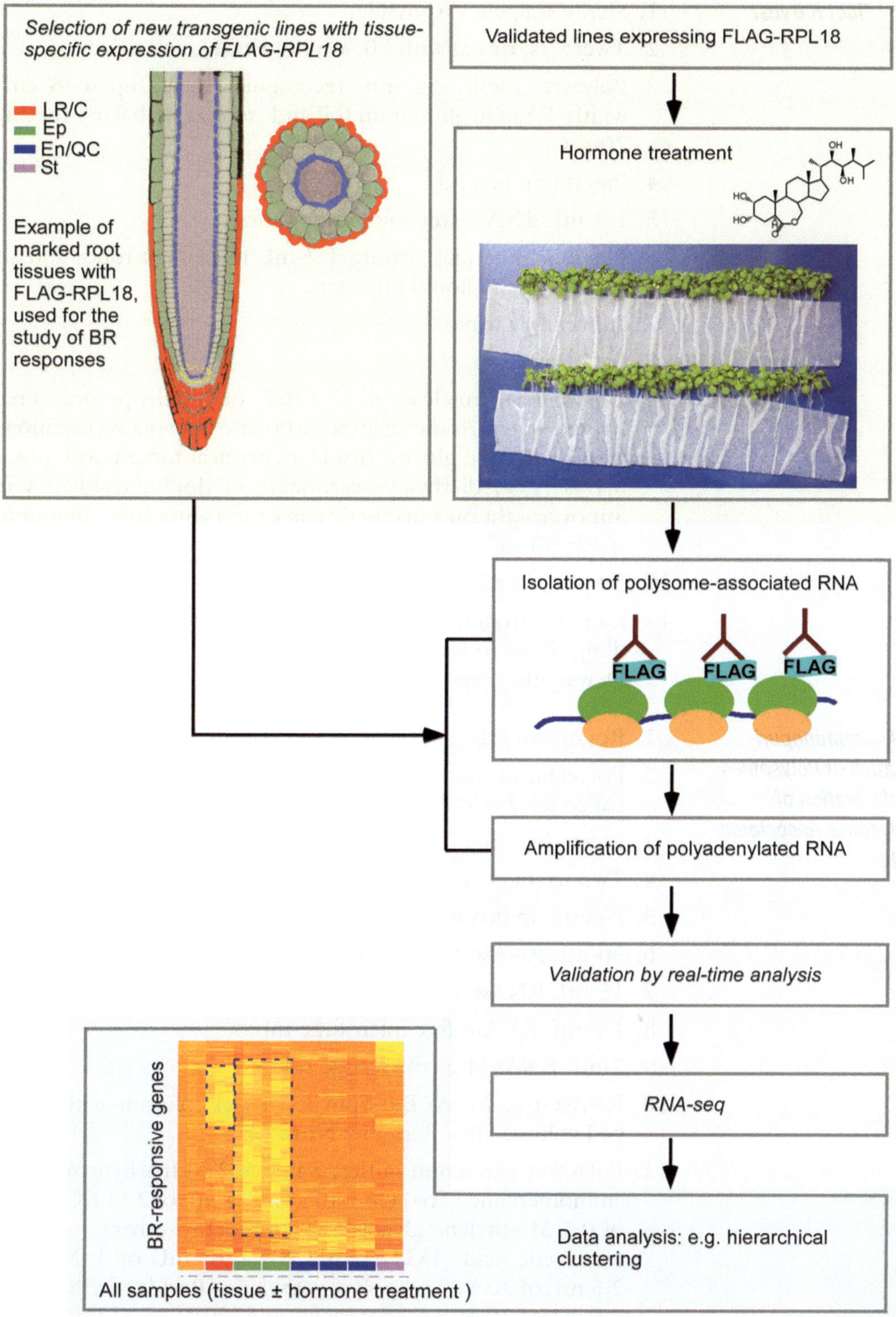

Fig. 1 Flowchart for quantitation of cell type-specific responses to BR by deep sequencing of polysome-associated polyadenylated RNA. Steps annotated in italics are briefly discussed. A detailed protocol is provided for all other steps.

2.2 Root Harvest

1. Sterile scalpels, disposable.
2. Tweezers, rinsed with 70 % (v/v) EtOH, and dried.
3. Polyester mesh, cut into rectangular sheets up to 6 cm in width. Wrap in aluminum foil and sterilize by baking for 2 h at 200 °C.
4. Pipette tip box lid.
5. 1.7-mL RNAse-free microfuge tubes.
6. Floating tube rack, fitting 1.7-mL microfuge tubes and suitable for use in liquid nitrogen.
7. Laboratory wipes.
8. Ice bucket.
9. Commercial nuclease-free water or diethylpyrocarbonate-treated water. As diethylpyrocarbonate is a volatile carcinogen, wear protective gloves. Inside a chemical fume hood, prepare 0.1 % (v/v) diethylpyrocarbonate in double-distilled water. Stir overnight on a magnetic stirrer and autoclave. Aliquot into sterile 50-mL Falcon tubes and store at room temperature.
10. 70 % (v/v) ethanol.
11. Liquid nitrogen. Wear appropriate protective gear when handling. Prior to use, pour into a 1-L capacity cryogenic storage Dewar and keep closed.

2.3 Immunopurification of Polysomes and Isolation of Polysome-Associated RNA

1. Rotational tube shaker accommodating 15-mL Falcon tubes.
2. Porcelain mortars and pestles, wrapped in aluminum foil and baked for 2 h at 200 °C.
3. Long tweezers.
4. Two ice buckets.
5. Pipette tip box lid.
6. 50-mL RNAse-free Falcon tubes.
7. 15-mL RNAse-free Falcon tubes.
8. 1.7-mL RNAse-free microfuge tubes.
9. 2-mL RNAse-free microfuge tubes.
10. RNAse-free Micro Bio-Spin P-30 Gel columns with 0.8 mL bed volume (Bio-Rad) (*see* **Note 4**)
11. Polysome extraction buffer: 5 mL of 2 M tris(hydroxymethyl) aminomethane (Tris) base (pH 9.0), 5 mL of 2 M KCl, 2.5 mL of 0.5 M ethylene glycol-bis(β-aminoethyl ether) *N,N,N′,N′*-tetraacetic acid (EGTA) (pH 8), 1.75 mL of 1 M $MgCl_2$, 2.5 mL of 20 % (v/v) polyoxyethylene 10 tridecyl ether (PTE), 2.5 mL of 20 % (v/v) detergent mix (20 % [w/v] Brij-35, 20 %

[v/v] Triton X-100, 20 % [v/v] Igepal CA 630, and 20 % [v/v] polyoxyethylenesorbitan [Tween 20]), 500 μL of 0.5 M dithiothreitol (DTT), 500 μL of 0.1 M phenylmethanesulfonyl fluoride (PMSF), 100 μL of 25 mg/mL cycloheximide, 50 μL of 50 mg/mL chloramphenicol, and 500 μL of 50 mg/mL heparin. Add water to reach a volume of 50 mL and keep on ice. This volume is sufficient for extracting polysomes from three harvested tissue samples. Prepare fresh on the day of the experiment (*see* **Note 5**).

12. Wash buffer: 5 mL of 2 M Tris base (pH 9.0), 5 mL of 2 M KCl, 2.5 mL of 0.5 M EGTA (pH 8), 1.75 mL of 1 M $MgCl_2$, 500 μL of 0.5 M DTT, 500 μL of 0.1 M PMSF, 100 μL of 25 mg/mL cycloheximide, and 50 μL of 50 mg/mL chloramphenicol. Add water to reach a volume of 50 mL and keep on ice. Approximately 100 mL of wash buffer is sufficient for precipitating polysomes from three harvested tissue samples. Prepare fresh on the day of the experiment.
13. Anti-FLAG M2 affinity agarose beads.
14. TRIzol Reagent.
15. Linear polyacrylamide carrier (25 mg/mL) for RNA precipitation.
16. Chloroform.
17. 100 % (v/v) isopropanol.
18. 80 % (v/v) ethanol, cooled for 5 min at −80 °C and stored on ice before use.
19. Commercial nuclease-free water or diethylpyrocarbonate-treated water (*see* Subheading 2.2, **step 9**).
20. Liquid nitrogen (*see* Subheading 2.2, **step 11**).

2.4 Amplification of Polysome-Associated Polyadenylated RNA

1. Vacuum centrifuge concentrator, suitable for 1.7-mL and 2-mL microfuge tubes.
2. Thermocycler with adjustable heated lid.
3. Heat block set at 55 °C, for cDNA and antisense RNA (aRNA) purification.
4. MessageAmpII aRNA amplification kit (Ambion) (*see* **Note 26**).
5. 0.2-mL RNAse-free PCR tubes (*see* **Note 6**).
6. 0.5-mL LoBind RNAse-free microfuge tubes (Eppendorf).
7. 1.7-mL RNAse-free microfuge tubes (*see* **Note 6**).
8. 100 % (v/v) ethanol.
9. Nuclease-free water.

3 Methods

3.1 Selection of Plant Material

As is typical for transformation events, the expression of FLAG-RPL18 will show variations between independent transformants and, as such, will impact on the enrichment of tissue-specific RNA with respect to nonspecific RNA (such as RNA presence in other tissues). To select the most suitable line for translatome profiling, proceed with at least three independent transformants, indicating a single T-DNA insertion by Mendelian segregation ratio on the antibiotic selection (*see* **Note 7**). In the T2 generation, immuno-purify the FLAG-tagged polysomes and isolate the polysome RNA (*see* Subheading 3.4). Amplify the isolated polysome RNA by in vitro transcription (*see* Subheading 3.5) and use 2 μL of aRNA for qRT-PCR (*see* **Note 8**). Quantify the fold RNA enrichment of known genes present in the tissue of interest (the endogenous gene corresponding to the promoter in use should be a good marker gene for that tissue), for instance, by comparing the expression of the marker gene between the transformant lines and a line with a broad expression pattern of FLAG-RPL18 (such as *p35S* promoter) (Fig. 2). A successful transformant line typically shows high RNA enrichment of the marker gene relative to the lines broadly expressing FLAG-RPL18, indicative of efficient tissue-specific capturing of the RNA. Proceed with the selected transformant until the homozygous stage (*see* **Note 9**).

3.2 Hormone Treatment

Arabidopsis thaliana (L.) Heynh. seed plating and BL treatment are done inside a laminar cabinet.

1. Surface sterilize *Arabidopsis* seeds with the dry method. Place 50 mg of seeds in an open 1.7-mL microfuge tube next to a glass beaker with 50 mL of bleach. Add 1 mL of 37 % (v/v) HCl in the beaker and immediately close the desiccator. Keep in the closed desiccator for at least 4 h, up to overnight.
2. Align two rows of seeds on each square petri plate with solid ½ MS medium by using a sterile pipette tip or an inoculation loop. Seal the plates with Parafilm and stratify in the dark for 48 h.
3. Place the plates vertically in the growth chamber (set to 22 °C and a desired light regime) and grow the seedlings until BL treatment.
4. Switch on the laminar cabinet and let run for 10–15 min.
5. Label 50-mL Falcon tubes (or 250-mL Erlenmeyer flasks) as BL (hormone treatment) or mock (control treatment).
6. Defrost the BL stock at room temperature and prepare a 100-nM working solution in liquid ½ MS (*see* **Note 10**). Flick the tube to mix and pipette to a small volume (0.5–1 mL) of ½

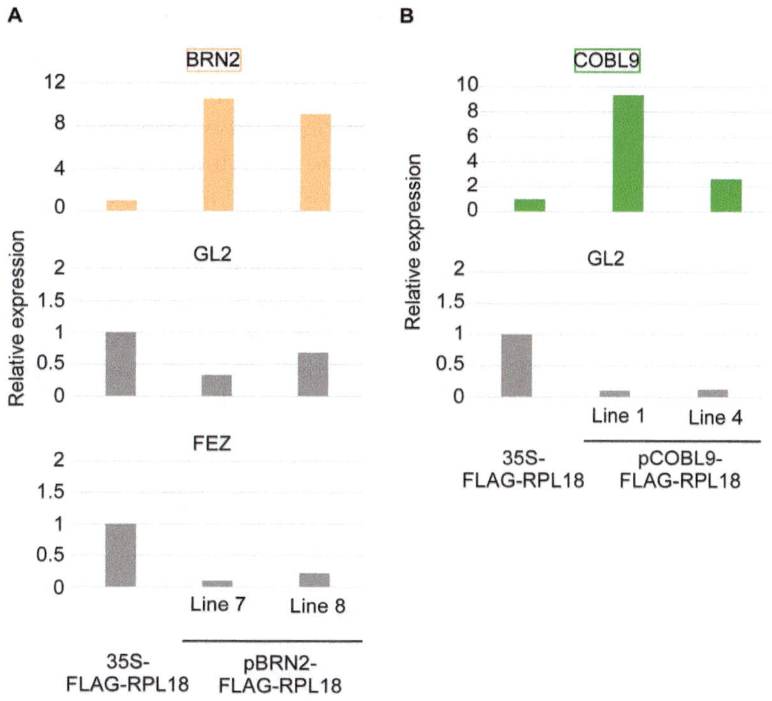

Fig. 2 Quantitative RT-PCR analysis comparing independent transgenic lines expressing tissue-specific FLAG-RPL18. Two independent transgenic lines (T2 generation) expressing FLAG-RPL18 under the pBRN2 promoter (**a**) and the pCOBL9 promoter (**b**) are compared to a line expressing FLAG-RPL18 under the 35S promoter (p35S-FLAG-RPL18). Note that the endogenous gene corresponding to each promoter (marked in *orange* and *green* in **a** and **b**, respectively) is enriched in the corresponding transgenic lines as compared to the p35S-FLAG-RPL18. Genes that are preferentially expressed in other tissues (marked in *gray*) are not enriched. Also note that the magnitude of enrichment differs between independent transgenic lines. For example, in panel **b**, line 1 is more suitable than line 4. One biological repetition is shown as an example. To normalize the variance among samples, the At5g15400 transcript level was used as an endogenous control.

MS in a 1.7-mL microfuge tube. Vortex briefly, spin down, and transfer with a pipette to the BL-labeled Falcon tube containing 90 % (v/v) of the final solution volume. Bring to the final volume and vortex vigorously. For treatments requiring 100 mL or more solution, add a small volume of ½ MS-dissolved BL into a graduated glass cylinder containing 90 % (v/v) of the final volume. Bring to the final volume. Close the cylinder firmly with a piece of Parafilm and mix thoroughly by inverting several times. Transfer to the BL-labeled Erlenmeyer flask.

7. Prepare a mock solution by adding an equal volume of 100 % (v/v) DMSO to the liquid ½ MS.

8. Take the petri plates with seedlings out of the growth chamber and remove the Parafilm seal (*see* **Note 11**).
9. Mark the treatment type (BL or mock) and the treatment duration on the top corner of the petri lid (*see* **Notes 12** and **13**). If the experimental sample consists of more than five plates (transgenic lines expressing the FLAG-tagged polysomes in a few cells), divide into sets of five plates and mark each with treatment type and number (1, 2, 3, etc.).
10. Align the plates in a single row. Remove the lids and set them aside. Starting from the first plate, pipette 1 mL of the treatment solution and purge in one continuous motion over the roots of both seedling rows. After treating all the plates, gently cover the roots with a precut 1-cm wipe strip without disturbing the shoots. Apply autoclaved wipes with ethanol-rinsed tweezers if treatment duration exceeds 8 h. Repeat with another 1 mL of the treatment solution and close the lids (*see* **Note 14**).
11. Allow the solution to absorb for 10 min and begin treating the next experimental sample. After 10 min, return the first experimental sample to the growth chamber and maintain in a vertical position until root harvest. Proceed in the same manner with each consecutive sample.

3.3 Root Harvest

Take special care to carry out all the steps under clean, RNase-free conditions (*see* **Note 15**).

1. Place an ice bucket on the laboratory bench. Prepare a cold cutting surface on ice by using a clean tip box lid. Press the lid into the ice (*see* **Note 16**).
2. Label 1.7-mL microfuge tubes with the sample name, treatment type, and treatment duration. Organize the tubes in a tube rack in the exact order of harvesting, starting from the shortest treatment and the first treated sample (*see* **Note 17**).
3. Position a Dewar filled with liquid nitrogen (*see* Subheading 2.2, **step 11**) next to the ice bucket. Open the first microfuge tube and insert into the floating rack. Drop the rack gently into liquid nitrogen without submerging the tube.
4. Take out the first sample from the growth chamber and spread the plates on a clean absorbent paper (*see* **Note 18**). If sets of plates per sample are harvested, take out the first treated set (marked as 1; *see* Subheading 3.2, **step 9**), while leaving the remaining sets in the growth chamber. The second set should be taken out only after harvesting of the first set is completed.
5. Discard the plate lids and gently remove the wipe strips with ethanol-rinsed tweezers. Place a mesh sheet on the cutting surface, and wet with nuclease-free water.

6. Collect the seedlings with tweezers and transfer to the mesh sheet without release. To enhance speed, harvest both seedling rows at once. With a clean scalpel, cut one-third of the entire root bundle, measuring from the root tips, and transfer into the nitrogen-chilled microfuge tube (*see* **Note 19**).
7. Close the microfuge tube when completed. Lift the rack and close the tube immediately. Drop the tube into nitrogen (*see* **Note 20**). Before harvesting the following sample, rinse tweezers and scalpel with ethanol and place a fresh mesh sheet on the cutting surface.
8. Store the harvested roots at −80 °C.

3.4 Immunopurification of FLAG-Tagged Polysomes and Isolation of Polysome-Associated RNA

All centrifugations are performed at 4 °C. Unless indicated otherwise, all steps are done on ice and inside a chemical fume hood.

1. Cool down a microcentrifuge (16,000 × *g*) and a low-speed centrifuge (4000 × *g*, for rotating 15-mL and 50-mL Falcon tubes) to 4 °C. Prepare the necessary buffers (*see* Subheading 2.3, **steps 11** and **12**).
2. Transfer the harvested roots frozen at −80 °C into a Dewar filled with liquid nitrogen (*see* Subheading 2.2, **step 11**).
3. Place two ice buckets in the chemical fume hood. In the first bucket, prepare a solid support for tissue grinding by pressing a tip box lid into the ice. Position a clean porcelain mortar on top of the support and fill with liquid nitrogen. Using long tweezers, take out a microfuge tube from the Dewar and empty the content into the mortar (*see* **Note 21**). Grind the tissue with a pestle until a fine white powder is obtained, and then add twice the volume of ice-cold polysome extraction buffer. Incubate on ice in the second bucket until completely thawed. If multiple samples are ground, put stickers with sample abbreviations next to the mortars to avoid confusion.
4. Transfer the tissue homogenate by pipetting into a 1.7-mL microfuge tube and centrifuge at 16,000 × *g* for 15 min.
5. Pipette the supernatant into a clean 1.7-mL microfuge tube and centrifuge for an additional 15 min at 16,000 × *g*. Pipette the purified supernatant into a 15-mL Falcon tube and set aside.
6. Prepare anti-FLAG M2 affinity agarose beads. Spin down the affinity gel at 4000 × *g* for 2 min in the low-speed centrifuge, by inserting the stock vial into a 50-mL Falcon tube. Gently rotate the vial to resuspend the beads and pipette 100 μL into a 1.7-mL microfuge tube (*see* **Note 22**). Centrifuge the tube at 8200 × *g* for 2 min. With a narrow pipette tip, carefully remove the resin without collecting the beads. Resuspend with 1.5 mL of wash buffer and centrifuge at 8200 × *g* for 2 min. Remove the supernatant and repeat the washing step with 1.5 mL of wash buffer.

7. Mix 100 μL of the washed beads with the supernatant (*see* Subheading 3.4, **step 5**). Bring the volume to 5 mL with the polysome extraction buffer.
8. Incubate on a shaker in a 4 °C refrigerator or a cold room by gentle rocking with approximately a 100-degree angle at 14 rpm for 2 h. The FLAG-tagged polysomes are now bound to the agarose beads.
9. Spin down the incubated beads in the low-speed centrifuge at 4000 × *g* for 2 min. Remove the supernatant without disturbing the beads. Bring the volume to 6 mL with the polysome extraction buffer. Incubate on the shaker by rocking at 14 rpm for 15 min.
10. Assemble a washing column. Sharply invert the Micro Bio-Spin P-30 Gel column to resuspend the gel. Completely remove and discard the gel. Rinse the column thoroughly with the wash buffer several times and snap off the bottom tip. Seat the column in a 2-mL microfuge tube and centrifuge at 8200 × *g* for 1 min to remove the buffer.
11. Spin down the beads incubated with the polysome extraction buffer at 8200 × *g* for 2 min. Remove the buffer.
12. Resuspend the beads with 500 μL of wash buffer by gentle pipetting up and down, and transfer into the washing column. Centrifuge at 8200 × *g* for 30 s and discard the flow-through. Add another 500 μL of wash buffer and repeat the centrifugation. Repeat this washing step two to three times (*see* **Note 23**).
13. Resuspend the beads with the wash buffer and transfer from the column into a clean 15-mL Falcon tube. Bring the volume to 6 mL with wash buffer and incubate on the shaker by rocking at 14 rpm for 15 min.
14. Spin down the beads incubated with the wash buffer at 8200 × *g* for 2 min. Remove the buffer and wash the beads in the column (*see* Subheading 3.4, **step 12**). Resuspend and transfer into a clean 15-mL Falcon tube. Incubate twice for 15 min with 6 mL of wash buffer, each followed by extensive washing of the column.
15. Transfer the beads to a clean 1.7-mL microfuge tube and discard the washing column. Centrifuge briefly at 8200 × *g* and completely remove the wash buffer with a narrow pipette tip.
16. Spin down the linear polyacrylamide carrier (*see* Subheading 2.3, **step 15**) and store on ice. Add 1 mL of TRIzol reagent and 2 μL of the carrier to the agarose beads and vortex vigorously. Incubate for 5 min at room temperature and centrifuge at 14,000 × *g* for 5 min.

17. Transfer the supernatant to a clean 1.7-mL microfuge tube and discard the agarose beads (*see* **Note 24**).
18. Add 200 μL of chloroform and vortex vigorously for at least 30 s. Incubate at room temperature for 10 min and centrifuge at 12,000 × *g* for 15 min. With a narrow 10-μL tip, carefully transfer the upper aqueous phase containing the polysome-associated RNA to a clean 1.7-mL microfuge tube. Repeat the extraction with another 200 μL of chloroform.
19. Add an equivalent volume (approximately 500 μL) of isopropanol to the aqueous phase and invert the tube several times.
20. Precipitate the polysome RNA overnight at −20 °C.
21. Centrifuge the precipitated RNA at 12,000 × *g* for 30 min.
22. Carefully remove the supernatant from the RNA pellet.
23. Wash the pellet by adding 1 mL of ice-cold 80 % (v/v) ethanol. Invert the tube several times and centrifuge at 12,000 × *g* for 30 min. Remove the supernatant and repeat the ethanol wash.
24. Completely remove the ethanol from the microfuge tube. After removal of the supernatant, centrifuge the tube at 7400 × *g* for 2 min. With a narrow 10-μL tip, remove the residual ethanol.
25. Air-dry the pellet for 3–5 min and resuspend in 15 μL of RNAse-free water by flicking the tube until fully dissolved. The isolated polysome-associated RNA can now be stored at −80 °C (*see* **Note 25**).

3.5 Amplification of Polyadenylated RNA

1. Heat 2 μL of the polysome-associated RNA (*see* Subheading 3.4) at 70 °C for 5 min.
2. Cool on ice.
3. Set up the first-strand cDNA synthesis reaction in a 0.2-mL PCR tube as follows (*see* **Note 26**): 2 μL of polysome-associated RNA (from Subheading 3.4), 2 μL of nuclease-free water, 0.2 μL of T7 oligo(dT) primer, 0.4 μL of 10× first-strand buffer, 0.8 μL of dNTP mix, 0.2 μL of RNAse inhibitor, and 0.2 μL of ArrayScript reverse transcriptase.
4. Mix thoroughly by pipetting up and down several times. Flick and spin down.
5. Preheat the lid of a thermocycler to 50 °C.
6. Incubate for 2 h at 42 °C (*see* **Note 27**).
7. Cool on ice.
8. Set up the second-strand cDNA synthesis reaction as follows: 12.6 μL of nuclease-free water, 4 μL of first strand cDNA reaction product (*see* Subheading 3.5, **step 3**), 2 μL of 10× second-strand buffer, 0.8 μL of dNTP mix, 0.4 μL of DNA polymerase, and 0.2 μL of RNAse H.

9. Mix thoroughly by pipetting up and down several times. Flick and spin down the tube.
10. Incubate for 2 h at 16 °C in a thermocycler with an open lid (*see* **Note 27**).
11. Purify the synthesized cDNA at room temperature. Add 80 μL of nuclease-free water and transfer to a clean 1.7-mL microfuge tube.
12. Bind the cDNA. Add 250 μL of cDNA-binding buffer and mix by pipetting up and down. Flick and spin down briefly. Transfer the entire volume to the center of the cDNA elution column. Centrifuge at 10,000 × *g* for 1 min and discard the flow-through.
13. Add 500 μL of wash buffer. Centrifuge at 10,000 × *g* for 1 min and discard the flow-through. Centrifuge for an additional 1 min to remove residual wash buffer.
14. Transfer the cDNA filter cartridge to a clean 2-mL cDNA elution tube.
15. Elute the cDNA. Add 9 μL of nuclease-free water preheated at 55 °C to the center of the elution column and incubate for 2 min. Centrifuge at 10,000*g* for 90 s. Repeat the elution step.
16. Remove and discard the cDNA filter cartridge. Place the cDNA elution tube in the vacuum centrifuge and concentrate to 6.4 μL (*see* **Note 28**).
17. Set up the in vitro transcription reaction to synthesize multiple copies of aRNA from the cDNA template. In a clean 0.2-mL PCR tube combine the following reagents: 6.4 μL of concentrated cDNA (*see* Subheading 3.5, **step 16**), 1.6 μL of ATP, 1.6 μL of CTP, 1.6 μL of GTP, 1.6 μL of UTP, 1.6 μL of T7 10× reaction buffer, and 1.6 μL of T7 enzyme mix.
18. Mix thoroughly by pipetting up and down several times. Flick and spin down the tube.
19. Preheat the lid of a thermocycler to 70 °C.
20. Incubate for 13 h at 37 °C (*see* **Note 27**).
21. Purify the synthesized aRNA at room temperature. Add 14 μL of nuclease-free water and transfer to a clean 1.7-mL microfuge tube.
22. Bind the aRNA. Add 105 μL of aRNA binding buffer followed by 75 μL of 100 % (v/v) ethanol. Immediately mix by pipetting up and down four to five times. Transfer the entire volume to the center of the cDNA elution column (*see* **Note 29**). Centrifuge at 10,000 × *g* for 1 min and discard the flow-through.

23. Add 500 μL of wash buffer. Centrifuge at 10,000 × *g* for 1 min and discard the flow-through. Centrifuge for an additional 1 min to remove residual wash buffer.
24. Transfer the cDNA filter cartridge to a clean 2-mL cDNA elution tube.
25. Elute the aRNA. Add 9 μL of nuclease-free water preheated at 55 °C to the center of the elution column. Incubate for 5 min in a 55 °C heat block. Centrifuge at 10,000 × *g* for 90 s. Repeat the elution step.
26. Store the purified aRNA at −80 °C (*see* **Note 8**).

3.6 RNA-seq Analysis and Bioinformatics

The aRNA can be handled by a unit service that sequences RNA-seq libraries on an Illumina HiSeq 2500 or similar next-generation sequencing machine. On average, 30 million single-end reads should be sequenced for each sample, with a read length of at least 50 base pairs. The steps below assume that the readers can install and run open source bioinformatics programs. Methods for the initial analysis of RNA-seq data are described, including an initial quality control of the reads, read alignment, read mapping to gene IDs, read count normalization, and determination of differentially expressed genes (DEG). These steps are routine in any RNA-seq analysis and alternative methods can be used, as long as the final outcome is a normalized table of read counts for each gene and sample. Finally, an Rscript is provided that takes a list of genes and a table of expression data (such as normalized read counts) and plots a hierarchically clustered heatmap. The list of genes can be derived from the DEG analysis, or from a gene ontology (GO) category or other functional annotation, or from genes identified in another study. Hierarchical clustering of the expression of a small set of genes of particular interest can give insight into the homogeneity (or conversely the heterogeneity) of these genes across a dataset.

1. Assess read quality with FastQC (http://www.bioinformatics.babraham.ac.uk/projects/fastqc/). Average read quality at each base pair should be above a phred quality score of 28 (*see* **Note 30**).
2. Align reads to the The Arabidopsis Information Resource (TAIR) 10 built of the *Arabidopsis thaliana* reference genome with a splice-aware alignment tool, such as Tophat2 [23], allowing, at most, three mismatches between read and reference. If Tophat2 is used, set the maximum and minimum intron sizes to 10 and 11,000 bp, respectively, and use Bowtie2 mapping option –b2-sensitive.
3. Create a read count table by associating read counts with annotated genes for each sample with HTseq-count in permissive intersection-non-empty mode (see http://www-huber.embl.de/users/anders/HTSeq/doc/count.html).

4. Use DESeq2 [24] both to normalize the read counts table according to the total read counts for each sample and to identify the DEGs.
5. Generate a Principal-component analysis (PCA) plot of the normalized data to qualitatively examine the correspondence between the known sample relationships and the observed expression similarities.
6. From the DESeq2 analysis, extract lists of DEGs with an adjusted *P* value ≤ 0.1 and a fold change ≥ 1.5 for each sample comparison of interest (*see* **Note 31**).
7. Use the Rscript makeHeatmapFromGeneListAndTable.R (available at https://github.com/ebartom/ClusteringExpressionData) to extract the normalized counts for a specific gene list of interest, hierarchically cluster them, and produce a heat map (Fig. 3a) (https://github.com/ebartom/ClusteringExpressionData/blob/master/wholeRoot.up3h.exp.heatmap.pdf). This tool can also be used to examine genes from an entirely different source in the data. For example, in Fig. 3b, genes from the longitudinal pattern 29 from the supplemental data [25] have been plotted in the normalized translatome data from [16] (Fig. 3b) (https://github.com/ebartom/ClusteringExpressionData/blob/master/longPattern29.tairOnly.exp.heatmap.pdf).

4 Notes

1. Other culture plates can be used, but square-shaped plates are more suitable for vertical positioning within the growth chamber. This orientation is important when *Arabidopsis* roots are examined, because it enables growth on the agar surface. The presence of the grid facilitates seed alignment. As an alternative, millimeter paper can be placed under plates without grids.
2. The composition is the same as of the solid ½ MS medium, but without addition of plant agar. With time, a yellow-white precipitate may appear at the bottom of the medium. Pipette with a sterile 25-mL serological pipette without disturbing the precipitate.
3. The working concentration of BL (100 nM) used in this protocol is extremely high. The aim is to probe the maximal perturbation of gene expression in response to the hormone. Different commercially available BRs are expected to act at different concentration ranges.
4. Other commercial columns are available, but most are not RNAse-free.

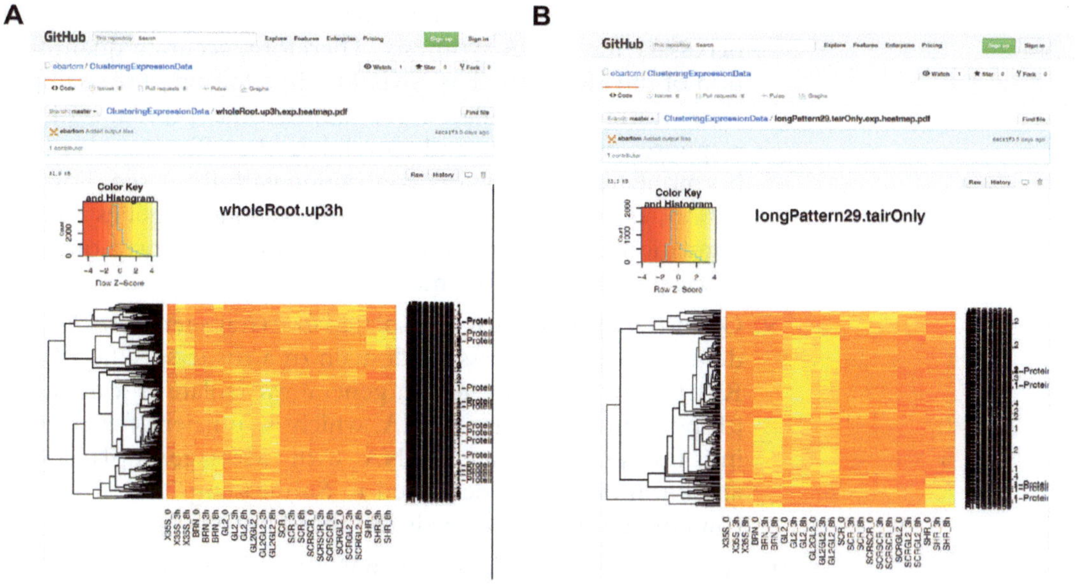

Fig. 3. Screen shots of results from an Rscript that takes a list of genes and a table of expression data and plots a hierarchically clustered heatmap. (**a**) Hierarchical clustering of genes induced in whole root (i.e., lines expressing p35S-FLAG-RPL18) after 3 h of BR treatment [16]. The expression levels across the entire dataset of each gene identified as differentially expressed is clustered and plotted. The color scale is adjusted for each gene independently, depending on the distribution of its expression levels. (**b**). Hierarchical clustering of BR-regulated genes, identified in representative dominant expression patterns along the longitudinal axis of the root (longitudinal pattern 29) [25], are plotted against the entire dataset.

5. Autoclave Tris base, EGTA, PTE, and the detergent mix stocks and keep at room temperature. Prior to autoclaving, heat the detergent mix in a water bath to 60 °C to dissolve Brij-35. Autoclave KCl and $MgCl_2$ stock solutions and after the first use, store at −20 °C to prevent bacterial contamination. Prepare DTT, cycloheximide (in ethanol), chloramphenicol (in ethanol), PMSF (in isopropanol), and heparin stocks as small aliquots, but do not autoclave. Store at −20 °C. Thaw all frozen stock solutions at room temperature before use.
6. If the thermocycler can accommodate 0.5-mL microfuge tubes, the entire protocol can be done in the LoBind tubes.
7. Single insertions allow faster establishment of the homozygous stage, especially in genetic crosses with FLAG-RPL18 lines.
8. The purified aRNA can be used as a template for cDNA library preparation or a qRT-PCR assay. In general, 2 μL of the aRNA product is sufficient for cDNA synthesis, so that the final aRNA volume (approximately 16 μL) can be used as a stock for several independent qRT-PCR assays, for instance, up to eight assays for abundant tissues, such as stele or endodermis. However, for transgenic lines in which FLAG-RPL18 is

expressed in a few cells, the entire final aRNA volume may be necessary for cDNA synthesis. Therefore, accurate quantification of the aRNA stock is advisable. In addition, if the samples were hormone-stimulated for translatome profiling of BR responses, we recommend to validate the effectiveness of the hormone treatment before continuing with RNA-seq. For validation, the behavior of known BR-responsive genes (downregulation or upregulation) can be used as a read-out of a successful temporal treatment.

9. Quantity of polysome RNA isolated from transgenic *Arabidopsis* depends on the expression domain of a tissue-specific FLAG-RPL18-driving promoter. In general, we isolate at least 3 ng/μL polysome-associated RNA, which is sufficient for amplifying the polyadenylated RNA with MessageAmpII aRNA amplification kit (as described in Subheading 3.5), from two rows of densely plated seeds (Subheading 2.1; Fig. 1).
10. Approximately 2 mL of 100 nM BL solution is sufficient to treat two rows of seedlings in the square petri plates. Return the BL stock solution to −20 °C immediately after use.
11. In the growth chamber, residual liquid often accumulates at the bottom of the plates with time. During treatment, prevent contact between any accumulated liquid and the seedlings, because it dilutes the treatment solutions. Discard the liquid onto a clean absorbent paper before applying the solutions.
12. BL treatment is always performed first, so position the plates with experimental samples intended for this treatment in the front rows of the growth chamber. Mock treatments are done immediately after their corresponding hormonal treatments. To avoid the effects of the circadian rhythm on gene expression, it is preferable to conduct the treatment at the same time of day.
13. The treatment length is determined based on the time interval within which the researcher wishes to analyze the transcript profile on the polyribosomes. Short treatments (30 min to 3 h) are adequate for capturing the early BR-responsive transcripts. In general, for profiling BR responses involved in meristem development, we use 6-day-old seedlings for treatments lasting up to 8 h.
14. When handling many plates for the same treatment time point (such as 3-h treatment or other), we recommend assistance from a colleague to enable exposure of all the samples to the hormone at approximately the same time of the day.
15. When DNA cloning procedures are used, remove bacterial cultures, DNA plasmids, and RNAse-containing reagents from the RNA workspace. Wipe pipettes with a tissue soaked in distilled water and leave to dry. Before starting with any of the

protocol sections, wipe all the working surfaces with 70 % (v/v) EtOH and use clean gloves.

16. Roots of *Arabidopsis* seedlings are semi-transparent. To facilitate the cutting, insert a piece of dark-colored plastic sheet beneath the lid.
17. It is essential to harvest the samples in the same order in which they were treated.
18. Precise timing is crucial for a successful temporal assessment of hormone-induced responses. Therefore, begin with the harvest exactly when the treatment time has expired. When harvesting many plates at the same time point, work with a colleague to harvest all the samples at approximately the same time. This point is especially important when harvesting closely spaced treatment durations (e.g., 30 min and 1 h).
19. After the cut has been made, significant changes in RNA composition will occur within a few minutes. Hence, immediately freeze the roots in nitrogen after cutting.
20. This step must be carried out quickly to prevent tissue thawing, because it compromises RNA integrity. Do not attempt to close the tube forcefully, because the built-up pressure can cause injuries. While closing the tube, dip the tube bottom in nitrogen several times to prevent thawing.
21. Rub the tube neck with your fingers until the cap snap opens, but do not apply force. Unload the roots into the mortar before nitrogen evaporation. Begin grinding only after full evaporation. Continue adding small portions of nitrogen until the fine powder is formed.
22. Beads do not pass through a regular pipette tip. Use clean, ethanol-rinsed scissors to cut 2 mm below the tip of the pipette tip. Also, beads tend to stick to the walls of the tips, the bottom of the washing column, and other equipment. Rinse all the surfaces with the appropriate buffer (indicated in each step), to recover as many beads as possible.
23. The column washing removes RNA molecules nonspecifically bound to the anti-FLAG M2 antibodies. As a control for the nonspecific RNA, incubate the agarose beads with wild-type root extract, i.e., not transformed with *FLAG-RPL18*. Subject 2 μL of aRNA (obtained as described in Subheading 3.5) to qRT-PCR. In the wild-type sample, the relative expression of the marker transcripts is extremely low, approximately 20–100-fold lower than the that of the transgenic lines. However, when a higher expression of the tested genes is detected in the wild-type sample, increase the number of column washes.
24. The supernatant contains the polysomes and the associated transcripts. It is very important to recover as much of the

supernatant as possible without collecting any beads. Note that both the beads and the supernatant have a similar color. If beads were accidentally collected, return the beads to the tube and centrifuge the sample again.

25. Avoid overdrying the pellet. If difficult to dissolve, spin down the microfuge tube and heat for 5 min at 55 °C in a heat block. Pipette up and down several times, flick, and spin down. Heat for an additional 5 min at 55 °C. Upon isolation, determine the integrity of the polysome RNA before proceeding with in vitro amplification (Subheading 3.5). We recommend BioAnalyzer instruments for accurate quantification. Polysome RNA should yield clear 25S and 18S rRNA peaks with RIN values in the nine to ten range.
26. cDNA synthesis and in vitro transcription of aRNA are done using the reagents provided in the MessageAmpII kit. Thaw all the frozen reagents of the MessageAmpII kit at room temperature and vortex thoroughly. Keep on ice until use. If many samples are amplified, prepare master mixes in 0.5-mL LoBind microfuge tubes with 5 % extra to compensate for pipetting errors.
27. Avoid thermocycler ramping. Place the tube in the sample block when it reaches the working temperature.
28. It takes roughly 8 min to concentrate the cDNA. After 8 min, check the volume of the cDNA droplet with a pipette set to 6.4 μL. If still above 6.4 μL, continue concentrating and rechecking the volume every minute. It is important not to overdry the droplet, because it may inhibit the in vitro transcription reaction (Subheading 3.5, **step 17**).
29. To avoid losing the small amounts of aRNA obtained by this protocol, we recommend to carry out the entire aRNA purification in the cDNA elution columns. If multiple samples are handled, perform aRNA purification separately for every sample. After binding and addition of the wash buffer (Subheading 3.5, **step 23**), each sample can wait until binding of all the samples is completed.
30. A Phred score is defined as $Q = -10 \log_{10} P$, where P is the probability of a base-calling error. For example, if the estimated probability that a base is called incorrectly is 1/1000, that base has a Phred quality score of 30.
31. Increasing the threshold for adjusted *P* values will increase the number of expected false positives in the results, whereas decreasing it may increase the number of false-negatives. An adjusted *P* value of 0.1 suggests 10 % false positives. The purpose of using a secondary threshold based on fold change is to eliminate small-scale expression changes, which may not be as biologically relevant as large changes.

Acknowledgments

This work was supported by grants received from the US-Israel Agricultural Research and Development Fund (IS-4246–09) and the Israel Science Foundation (592/13).

References

1. Bailey-Serres J (2013) Microgenomics: genome-scale, cell-specific monitoring of multiple gene regulation tiers. Annu Rev Plant Biol 64: 293–325
2. Birnbaum K, Shasha DE, Wang JY et al (2003) A gene expression map of the *Arabidopsis* root. Science 302:1956–1960
3. Iyer-Pascuzzi AS, Benfey PN (2010) Fluorescence-activated cell sorting in plant developmental biology. In: Hennig L, Köhler C (eds) Plant developmental biology. Methods in Molecular Biology, vol 655. Springer, New York, pp 313–319
4. Deal RB, Henikoff S (2011) The INTACT method for cell type-specific gene expression and chromatin profiling in *Arabidopsis thaliana*. Nat Protoc 6:56–68
5. Deal RB, Henikoff S (2010) A simple method for gene expression and chromatin profiling of individual cell types within a tissue. Dev Cell 18:1030–1040
6. Zanetti ME, Chang I-F, Gong F et al (2005) Immunopurification of polyribosomal complexes of Arabidopsis for global analysis of gene expression. Plant Physiol 138:624–635
7. Mustroph A, Zanetti ME, Jang CJH et al (2009) Profiling translatomes of discrete cell populations resolves altered cellular priorities during hypoxia in *Arabidopsis*. Proc Natl Acad Sci USA 106:18843–18848
8. Bargmann BO, Vanneste S, Krouk G et al (2013) A map of cell type-specific auxin responses. Mol Syst Biol 9:688
9. Bruex A, Kainkaryam RM, Wieckowski Y, Kang YH, Bernhardt C, Xia Y, Zheng X, Wang JY, Lee MM, Benfey P, Woolf PJ, Schiefelbein J (2012) A gene regulatory network for root epidermis cell differentiation in Arabidopsis. PLoS Genet 8:e1002446
10. Vilarrasa-Blasi J, Gonzalez-García MP, Frigola D, Fàbregas N, Alexiou KG, López-Bigas N, Rivas S, Jauneau A, Lohmann JU, Benfey PN, Ibañes M, Caño-Delgado AI (2014) Regulation of plant stem cell quiescence by a brassinosteroid signaling module. Dev Cell 30:36–47
11. Richter JD, Coller J (2015) Pausing on polyribosomes: make way for elongation in translational control. Cell 163:292–300
12. Ingolia NT, Ghaemmaghami S, Newman JRS et al (2009) Genome-wide analysis in vivo of translation with nucleotide resolution using ribosome profiling. Science 324:218–223
13. Ribeiro DM, Araújo WL, Fernie AR et al (2012) Translatome and metabolome effects triggered by gibberellins during rosette growth in *Arabidopsis*. J Exp Bot 63:2769–2786
14. Wang Y, Jiao Y (2014) Translating ribosome affinity purification (TRAP) for cell-specific translation profiling in developing flowers. In: Reichmann JL, Wellmer F (eds) Flower Development. Methods in Molecular Biology, vol 1110. Springer, New York, pp 323–328
15. Juntawong P, Girke T, Bazin J et al (2014) Translational dynamics revealed by genome-wide profiling of ribosome footprints in *Arabidopsis*. Proc Natl Acad Sci USA 111: E203–E212
16. Vragović K, Sela A, Friedlander-Shani L et al (2015) Translatome analyses capture of opposing tissue-specific brassinosteroid signals orchestrating root meristem differentiation. Proc Natl Acad Sci USA 112:923–928
17. Merchante C, Brumos J, Yun J et al (2015) Gene-specific translation regulation mediated by the hormone-signaling molecule EIN2. Cell 163:684–697
18. Singh AP, Savaldi-Goldstein S (2015) Growth control: brassinosteroid activity gets context. J Exp Bot 66:1123–1132
19. Hacham Y, Holland N, Butterfield C et al (2011) Brassinosteroid perception in the epidermis controls root meristem size. Development 138:839–848
20. Yoo SY, Bomblies K, Yoo SK et al (2005) The 35S promoter used in a selectable marker gene of a plant transformation vector affects the expression of the transgene. Planta 221:523–530
21. Wang X, Chory J (2006) Brassinosteroids regulate dissociation of BKI1, a negative regulator of BRI1 signaling, from the plasma membrane. Science 313:1118–1122

22. Yin Y, Wang Z-Y, Mora-Garcia S et al (2002) BES1 accumulates in the nucleus in response to brassinosteroids to regulate gene expression and promote stem elongation. Cell 109:181–191
23. Kim D, Pertea G, Trapnell C et al (2013) TopHat2: accurate alignment of transcriptomes in the presence of insertions, deletions and gene fusions. Genome Biol 14:R36
24. Love MI, Huber W, Anders S (2014) Moderated estimation of fold change and dispersion for RNA-seq data with DESeq2. Genome Biol 15:550
25. Brady SM, Orlando DA, Lee J-Y et al (2007) A high-resolution root spatiotemporal map reveals dominant expression patterns. Science 318:801–806

Chapter 9

Methods for Modeling Brassinosteroid-Mediated Signaling in Plant Development

David Frigola, Ana I. Caño-Delgado, and Marta Ibañes

Abstract

Mathematical modeling of biological processes is a useful tool to draw conclusions that are contained in the data, but not directly reachable, as well as to make predictions and select the most efficient follow-up experiments. Here we outline a method to model systems of a few proteins that interact transcriptionally and/or posttranscriptionally, by representing the system as Ordinary Differential Equations and to study the model dynamics and stationary states. We exemplify this method by focusing on the regulation by the brassinosteroid (BR) signaling component BRASSINOSTEROID INSENSITIVE1 ETHYL METHYL SULFONATE SUPPRESSOR1 (BES1) of BRAVO, a quiescence-regulating transcription factor expressed in the quiescent cells of *Arabidopsis thaliana* roots. The method to extract the stationary states and the dynamics is provided as a Mathematica code and requires basic knowledge of the Mathematica software to be executed.

Key words Brassinosteroids, Transcriptional circuit, ODEs, Steady states, Dynamical systems, Modeling

1 Introduction

Since the last decade, scientists with backgrounds in sciences, such as physics or mathematics, have started to collaborate more frequently with experimental biologists with the aim to use mathematical modeling to facilitate gathering conclusions from experimental data. These models, which are a simplification of the biological systems, can enable to examine relationships that cannot be directly observed in experiments as well as to make predictions that do not derive obviously from the data.

Several historical and recent examples exist of fruitful application of mathematical and computational modeling to experimental data, many of which in unicellular organisms or the animal kingdom [1–3], that explain phenomena that range from protein production dynamics to pattern formation, but also in different plant developmental aspects, such as the integration of signals to regulate

Eugenia Russinova and Ana I. Caño-Delgado (eds.), *Brassinosteroids: Methods and Protocols*, Methods in Molecular Biology, vol. 1564, DOI 10.1007/978-1-4939-6813-8_9, © Springer Science+Business Media LLC 2017

division in the collumella [4], the relationship of auxin and the root quiescent center [5] and of auxin with cytokinins in the root [6], the effect of brassinosteroids (BRs) on root growth [7], or even sepal patterning [8], to name a few.

Of the many approaches for mechanistic models, each has its strengths and drawbacks. They range from descriptions that include as many elements of the system as possible to only a few representative or even phenomenological elements, or purely analytical to purely simulation-based computational methods. In all cases, it is essential to choose well the system to describe and to which level, because there is always a degree of idealization [9].

Here we restrict ourselves to methods that concentrate on the choice of minimal elements to explain a phenomenon and analyze them with mathematical tools. Excellent general tutorials on this type of descriptions are available [10–12]. In the case described in this chapter, the dynamics of the selected minimal elements are defined by a set of Ordinary Differential Equations (ODEs), which, for instance, can be obtained from a description of the dynamics in terms of chemical reactions. These ODEs contain information on the steady state and the dynamics of the system and can be manipulated to check whether the initial assumptions and simplifications are compatible with information gathered from mutants, treatments, and other experiments, as well as to obtain predictions about experiments yet to be performed, directing the experimental effort that can discriminate among differing assumptions or falsify nonintuitive predictions of the model.

Previously, we have applied this type of modeling to a small transcriptional circuit under BR signaling control in *Arabidopsis thaliana* roots [13], in which the transcription factor BRAVO, an R2R3-MYB transcription factor, MYB56 (At5g17800) is presented that is a BR signaling target and is expressed in the quiescent centre of *Arabidopsis* roots. From several types of experiments [13], it was possible to find that BRAVO was regulated by the protein BRASSINOSTEROID INSENSITIVE1 ETHYL METHYL SULFONATE SUPPRESSOR1 (BES1) [14, 15] upon BR signaling and to set a BES1-BRAVO regulation circuit module [13]. This circuit consists of BRAVO self-activation and molecular sequestration by its transcriptional inhibitor BES1 (in its active, dephosphorylated form) [13]. Mathematical modeling of each of these two types of elements, i.e., self-activation on the one hand, and sequestration by the transcriptional inhibitor, on the other hand, had indicated that they can provide bistable and ultrasensitive responses [16–18]. The modeling of the BES1-BRAVO module pinpointed that the stationary levels of BES1 and BRAVO could be opposed and that BR signaling controlled the strongest one [13]. In this chapter, we make use of a slightly simplified version of this model dynamics and provide a Mathematica [19] code to analyze its stationary solutions and dynamics. The simplification done here affects the dynamics, but not the stationary solutions of the full model [13].

The method described aims at enabling the reader to calculate the dynamics of this BES1-BRAVO circuit by running the code and to apply the analysis to different sets of parameters or conditions by changing the code appropriately and minimally. Therefore, this analysis can be performed with only a basic knowledge of the Mathematica software [19]. The method also includes an initial part of model building (*see* Subheading 3.1) and a part of setting testable predictions (*see* Subheading 3.5). These two parts are very different from the remaining ones, because they cannot be simplified in a running code. Yet, with enough knowledge of mathematics and the Mathematica programming, the code we provide could be extended for use in other models of regulatory circuits involving other molecules by setting the appropriate equations to be analyzed and integrated and to set predictions. Therefore, Subheadings 3.1 and 3.5 are not to be pursued step by step, but are an introduction, because they are a crucial part of the modeling.

To obtain the same type of results extracted by our code, various software packages can be used, preferably with symbolic algebra capabilities, e.g., a good alternative is Maple, and free options are Maxima or Sympy in Python, but numerical computation programs can also be used, such as Matlab and Octave). Some of the software packages compatible with the popular Systems Biology Markup Language are also suitable for part of the type of analysis we focus on. Furthermore, even when Mathematica is used, there are several manners the code can be written to obtain the results. Instead of discussing these multiple ways, we concentrate on one in particular to aid understanding the code provided.

2 Materials

Mathematica software [19].

3 Methods

3.1 Building the Model

1. Decide on the questions you want to address. For the following steps to work in an optimal manner, do not select a study too large in scope, but be very specific either in its components or aspects to probe. Originally, we wanted to study how the regulation from BRs on the quiescent center division interacted with that of the protein of interest BRAVO [13]. For this reason, the main elements in this model were the protein BRAVO and its regulation by BR signaling through BES1 [13]. Consider also which are the minimum crucial elements required to describe and understand the problem. This critical

step requires both knowledge and intuition [9]. Take into account which time scales differ in the important parts and processes of the system and which processes can be simplified as an effective process. In our model, all BR signals were grouped together as the BES1 dephosphorylation [13], in other words, perception of the hormone at the plasma membrane by a receptor, among others, was simplified and represented only by the active and inactive forms of BES1. Additionally, we did not detail the processes of transcription and translation, disregarding the messenger RNA and its processing. It is worth mentioning that by these simplifications it is necessary to take care that the aspects of interest in the problem are represented in the model, either as variables (such as chemical species) or as processes (such as specific reaction rates). Consider also that sometimes not all experimental data can be included in this first model building.

2. Write the dynamical equations that describe the rate of changes of the molecular concentrations of the elements selected by considering the transitions and chemical reactions involved in the model according to the data and assuming that the mass action law holds. As an example, we detail the formulation of the dynamical equation for BRAVO, which we represent as variable $\tilde{M}$. This equation gives the transcriptional production of BRAVO, its dimerization dynamics with active BES1 (BES1d, defined by variable $\widetilde{B_d}$) and a decay rate, which we assume to be proportional to the BRAVO concentration (with a degradation rate r_M) [13]:

$$\frac{d\tilde{M}}{dT} = P\left(\tilde{M},\widetilde{B_d}\right) - q_{D+}\tilde{M}\widetilde{B_d} + q_{D-}\tilde{C} - r_M\tilde{M}$$

where T stands for time and the production is given by the function $P\left(\tilde{M},\widetilde{B_d}\right)$ that depends on BRAVO itself and on BES1d. Dimerization and dedimerization rates are q_{D+} and q_{D-} respectively. To define the production function, we considered that there were distinct binding sites in the DNA for BRAVO and for BES1 and assumed that they did not interact [13]. The production function is composed of four additive terms: the production rate when neither BRAVO nor active BES1 are bound to the DNA (defined as constant a per unbound DNA), the production rate when only BRAVO is bound to the DNA (with rate $\varepsilon_1 a$, with $\varepsilon_1 > 1$ because it is activation per bound DNA to BRAVO and unbound DNA to BES1), the production rate when only BES1d is bound to DNA ($\varepsilon_2 a$, with $\varepsilon_2 < 1$ because it inhibits per unbound DNA to BRAVO and bound DNA to BES1), and the production rate when both BRAVO and BES1d are bound ($\varepsilon_3 a$, with $\varepsilon_3 \approx 1$ as extracted from empirical data per bound DNA to BRAVO and bound DNA to BES1d) [13]. We refer the reader to see the full production function in terms of the free DNA and the complexes of DNA bound to either BRAVO, BES1d, or both (*see* [13]). To define the production function, we use all these elements and add a simplification by taking into account that DNA dynamics are much faster than the remainder of the system variable dynamics, allowing us to find the equilibrium concentrations of the binding sites and to make the analysis more manageable. When taking into account that the amount of binding sites does not change over time and that they are in equilibrium with their binding molecules, the production function becomes:

$$P\left(\tilde{M},\widetilde{B_d}\right) = aD_{tot1}D_{tot2}\frac{1+\varepsilon_1 K_M\tilde{M}+\varepsilon_2 K_B\tilde{B}+\varepsilon_3 K_M\tilde{M}K_B\tilde{B}}{\left(1+K_M\tilde{M}\right)\left(1+K_B\tilde{B}\right)}$$

where the equilibrium binding constants of BRAVO and BES1d to DNA are K_M and K_B, respectively, and D_{tot1} and D_{tot2} are the total concentrations of binding sites for BRAVO and BES1d, respectively. For formulation of this type of production functions, see [11]. With these considerations, the model results in the following four equations that describe the rate of change of BRAVO ($\tilde{M}$), BES1d ($\widetilde{B_d}$), inactive BES1 ($\tilde{B}$), and the dimer BRAVO-BES1d ($\tilde{C}$):

$$d\tilde{M}/dT = aD_{tot1}D_{tot2}\frac{1+\varepsilon_1 K_M\tilde{M}+\varepsilon_2 K_B\tilde{B}+\varepsilon_3 K_M\tilde{M}K_B\tilde{B}}{\left(1+K_M\tilde{M}\right)\left(1+K_B\tilde{B}\right)} - q_{D+}\tilde{M}\widetilde{B_d} + q_{D-}\tilde{C} - r_M\tilde{M}$$

$$d\widetilde{B_d}/dT = -q_{D+}\tilde{M}\widetilde{B_d} + q_{D-}\tilde{C} - q_{p+}\widetilde{B_d} + q_{p-}\tilde{B} - r_{Bd}\widetilde{B_d}$$

$$d\tilde{B}/dT = b + q_{p+}\widetilde{B_d} - q_{p-}\tilde{B} - r_B\tilde{B}$$

$$d\tilde{C}/dT = q_{D+}\tilde{M}\widetilde{B_d} - q_{D-}\tilde{C} - r_C\tilde{C}$$

where all decays are assumed to be proportional to the concentration (with rate r_X being the X variable), and the phosphorylation and dephosphorylation rates are q_{p+} and q_{p-}. Mathematically, it is known that the range of qualitative dynamics driven by a model may not depend on all the parameters, but on their combinations. Hence, once the model is defined, it is useful to reduce the number of parameters without altering the type of model-driven solutions. The parameters are reduced by defining variables without dimensions that can be converted into one with dimensions by multiplying them through the appropriately chosen factor. Following this procedure, we define time as t times T_0 ($T = tT_0$). Note that both T and T_0 have time dimensions, whereas t is just a unitless number. We also define the concentration of BRAVO as a number M of times the concentration value C_0, and correspondingly for all the other concentrations ($\tilde{M} = MC_0$, $\tilde{B} = BC_0$, $\widetilde{B_d} = B_dC_0$, $\tilde{C} = CC_0$). We chose $T_0 = 1/d_m$ and $C_0 = D_{tot1}$. In other words, all time scales are rescaled to the characteristic degradation time of BRAVO and all concentrations to the concentration of the BRAVO DNA-binding sites. Introduction of these definitions in the above system of equations results in the following set of four equations that have one less effective parameter than above (*see* **Note 1**):

$$\frac{dM}{dt} = \alpha \frac{1 + \varepsilon_1 K_M M + \varepsilon_2 K_B B + \varepsilon_3 K_M M K_B B}{(1 + K_M M)(1 + K_B B)} - k_{D+} M B_d + k_{D-} C - M \quad (1)$$

$$\frac{dB_d}{dt} = -k_{D+} M B_d + k_{D-} C - k_{p+} B_d + k_{p-} B - d_{Bd} B_d \quad (2)$$

$$\frac{dB}{dt} = \beta + k_{p+} B_d - k_{p-} B - d_B B \quad (3)$$

$$\frac{dC}{dt} = k_{D+} M B_d - k_{D-} C - d_C C \quad (4)$$

with $\alpha = aD_{tot2}/d_M$, $\beta = b/(D_{tot1} d_M)$, $K_M = k_M D_{tot1}$, $K_B = k_B D_{tot1}$, $k_{X\pm} = q_{X\pm}/d_M$ except for $k_{D+} = q_{D+} D_{tot1}/d_M$ and $d_X = r_X/d_M$ and X stands for all the subscripts. This set of four equations will be analyzed in the Mathematica code provided below.

3. For each parameter in the equations, determine a reasonable physiological range and use the time scales at which the different physiological processes happen. Note that in the case of effective parameters, compounded as a representation of more than one process, this range might not be straightforward. In the code, we provide a set of default parameter values (*see* **Note 2**).

   The initial code reads:

```
(***Code for initialization and parameter values***)
Off[Solve::ratnz];(*This option eliminates a specific warning that appears in our calculations, but does not affect the results. Eliminate it if you want to see all warnings*)
(*We use the same notation as in the text, unless specified. However, subscripts in the parameters and variables in the text appear herein as regular text*)
(*M=BRAVO (for MYB56), Bd=dephosphorylated, active BES, B=phosphorylated, inactive BES, Compl=BES1-BRAVO heterodimer*)
(*In all the code, X' stands for the time derivative of variable X*)
Dir = NotebookDirectory[];
SetDirectory[Dir];
Clear[M];Clear[Bd];Clear[B];Clear[Compl];
M>=0&&M∈Reals;
Bd>=0&&Bd∈Reals;
B>=0&&B∈Reals;
Compl>=0&&Compl∈Reals;
$MaxExtraPrecision=200;
```

```
(*Adimensional parameters from the non-dimensional model
of Eqs. 1–4 Default values*)
    KM=70.00;
    KB=80.00;
    kDplus=1000.0;
    kDminus=10.0;
    β =150;α=90;
    dB=1.0;dBd=0.1;dC=1.0;
    ε1=3.9;ε2=0.07;ε3=1.3;
    kPminus=0.5;
    kPplus=0.1;
    Print["******Parameters initialized or changed******"]
```

3.2 Computing and Representing Nullclines

The next steps are done through the Mathematica code that is provided at the end of this Section.

1. Compute the nullclines of M and B_d in the space of variables M and B_d. For a definition of the nullclines, *see* **Note 3**. Because the system has four variables and the representation desired of the nullclines is in a two-dimensional space, the code proposes to assume the rest of the variables to be in instantaneous equilibrium with M and B_d, by setting the time derivatives of these variables to zero and solving the corresponding equations (in the code provided with function Solve). The code uses a For loop to explore the M and B_d values that set their time derivatives to zero, saves these values, and plots them with the ListLogLogPlot function (other functions, such as ListPlot or ListLogPlot could also be used).
2. Once the nullclines are plotted (Fig. 1), observe their shape. Note how changes in the shape of the nullclines could modify the posit ion of their crossing points, which are

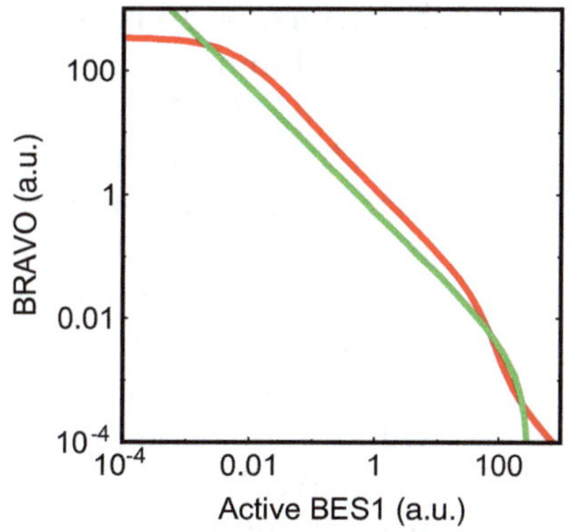

Fig. 1 Nullclines for active BES1 (*green*) and BRAVO (*red*) for the default parameter values ($d_C = 1$). There are three crossing points of the nullclines, which are fixed points. The *color code* used is the same as in ref. [13]

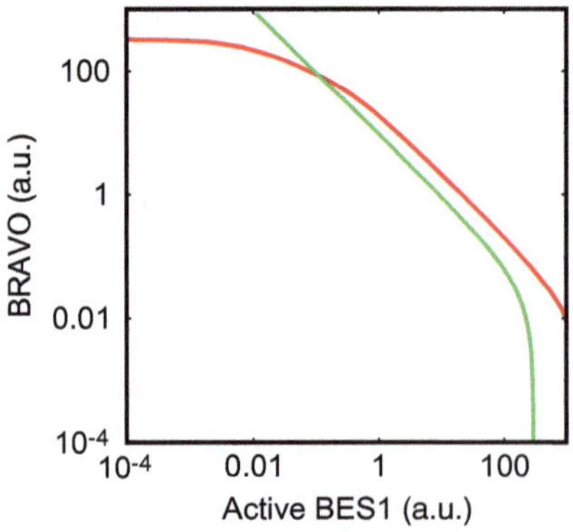

Fig. 2 Nullclines for active BES1 (*green*) and BRAVO (*red*) for $d_C = 0.05$ and the remaining parameters at their default value. There is only one single fixed point. Compared to Fig. 1, d_C is decreased and the shape of the nullclines has changed, so that the existence of three fixed points is impeded

called fixed points and correspond to stable or unstable stationary states.

3. Select a parameter and change its value and run the code again for computing and plotting nullclines. Look how the nullclines change, enabling to look for different qualitative behaviors, like those arising when the number of crossing points of the nullclines changes. For instance, with the provided code, change the value of parameter d_C and note how decreasing it to $d_C = 0.05$ makes the nullclines cross at a single point (Fig. 2). Repeat for different selected parameters.

The Mathematica code for computing the nullclines reads:

```
(***Nullclines***)
$MaxExtraPrecision=20000;
(*We can change parameters here if we want to see different behaviors, or go back to the first block and redefine them*)
(*kPminus=1.5;*)
(*dC=2.0;*)
(*There is a common method to draw nullclines in Mathematica that involves drawing the contours of M'=0 and B'=0, but it does not allow the use of log scales without workarounds. Instead, we manually explore many values of Bd and solve M'(Bd)=0 for each of them, and explore values of M and solve B'(M)=0*)
BravoNullcline={};Bes1Nullcline={} ;(*We initialize lists that will store the computed points of the nullcline*)
(*We do a loop in which we will solve each of the nullclines. As we only want to analyze two variables at the same time (M and Bd), we will set B and the Compl in equilibrium*)
For[valuex=0.00001,valuex<100000,valuex=valuex*1.5,
```

```
Clear[M];Clear[Bd];Clear[B];Clear[Compl];
(*First we set values for M and find the solutions of B'(M)=0*)
M= valuex;(*values of M*)
B'=β+kPplus*Bd-kPminus*B-dB*B; (*Sets the dynamical equation for phosphorilated (inactive) BES1 (Eq. 3)*)
Compl'=kDplus*Bd*M-kDminus*Compl-dC*Compl;(*Sets the dynamical equation for BRAVO-BES1 Compl (Eq. 4)*)
(*Now we set and solve the condition of B and Compl to be in equilibrium. CompST and BST are these solutions*)
CompST = Compl/.Solve[Compl'==0 && Compl>=0,Compl,Reals][[1]];
BST = B /.Solve[B'==0 && B>=0,B,Reals][[1]];
(*Now the nullcline Bd'(M)=0 is computed. B1 is the solution*)
Bd'=kPminus*BST-kPplus*Bd+kDminus*CompST-kDplus*M*Bd-dBd*Bd;
(*Dynamical equation for dephosphorylated (active) BES1 (Eq. 2)*)
B1=Bd/.Solve[{Bd'==0 && Bd>=0},Bd,Reals][[1]];
AppendTo[Bes1Nullcline,{B1,M}];(*We save the values of the nullcline, Bd and M, into a list Bes1Nullcline*)
(*We then set values for Bd and find the solutions of M'(Bd)=0*)
Clear[M];Clear[Bd];Clear[B];Clear[Compl];
Bd= valuex; (*values of Bd*)
B'=β+kPplus*Bd-kPminus*B-dB*B;
(*Sets the dynamical equation for phosphorylated (inactive) BES1 (Eq. 3)*)
Compl'=kDplus*Bd*M-kDminus*Compl-dC*Compl;
(*Sets the dynamical equation for BRAVO-BES1 Compl (Eq. 4)*)
(*Now we set and solve the condition of B and Compl to be in equilibrium. CompST and BST are these solutions*)
CompST = Compl/.Solve[Compl'==0 && Compl>=0,Compl,Reals][[1]];
BST = B/.Solve[B'==0 && B>=0,B,Reals][[1]];
(*Indeed BST is not required, since M does not explicitly depend on it*)
(*Production term for M, i.e., BRAVO*)
prodM=α*(1+ε1*KM*M+ε2*KB*Bd+ε3*KM*M*KB*Bd)/((1+KB*Bd) (1+KM*M));
(*Now the nullcline M'(Bd)=0 is computed. M1 is the solution*)
M'=prodM+kDminus*CompST-kDplus*M*Bd -M;
(*Dynamical equation for BRAVO (Eq. 1)*)
M1=M/.Solve[{M'==0 && M>=0},M][[1]]; AppendTo[BravoNullcline,{Bd,M1}]]
(*We plot all the computed results*)
```

```
range = {{0.0001,1000},{0.0001,1000}};
LogRange={{Log[0.0001],Log[1000]},{Log[0.0001],
Log[1000]}};
NullclinesPlot=ListLogLogPlot[{BravoNullcline,Bes1Nullcl
ine },PlotStyle->{Directive[Thick,Red],Directive[Thick,Green]},F
rame->True,FrameLabel->{"Active BES1(a.u.)","BRAVO(a.u.)"},
Joined->True,FrameStyle->Thick,LabelStyle-
>Directive[FontSize->16],ImageSize->300,AspectRatio-
>1,PlotRange->range,PlotLegends->{"BRAVO Nullc.","BES1
Nullc."}]
```

3.3 Stationary States and Bifurcation Diagrams

The next steps are done through the Mathematica code that is provided at the end of this subsection (*see* **Note 4**).

1. Compute the stationary states of the system. To this end, define again the dynamical equations of the system, equate them to zero, and solve them with the Nsolve or Solve functions (in the code provided this is done with Solve). The code provided also selects the functions that are real and positive for all the variables by means of the Select function. Note that this part of the code when run does not return any information written in the screen.
2. Select a parameter to be modified and to analyze how the stationary solutions and their number change accordingly. In the code, the chosen parameter is k_{p-}, but it can be replaced by o any other one. The For loop changes the value of the selected parameter over a range of values and the stationary solutions are computed for each value. The code returns a plot of these

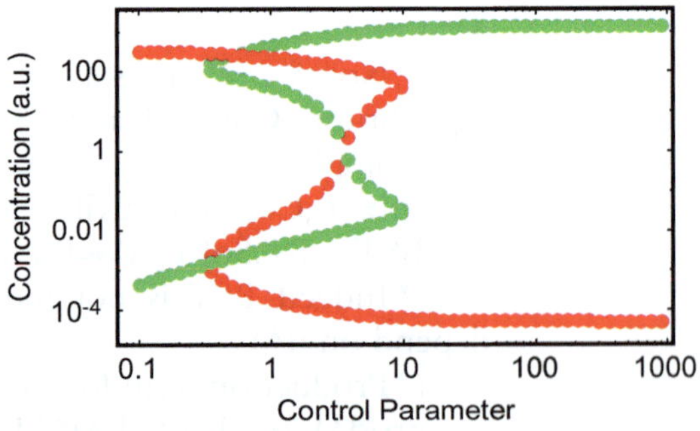

Fig. 3 Response curve/bifurcation diagram for active BES1 (*green*) and BRAVO (*red*) when the control parameter is the dephosphorylation rate k_{P-}, and the remaining parameters are at their default values. There is a range of k_{P-} values for which there are several solutions under the same conditions. At low k_{P-} the solution is of high BRAVO and low active BES1, whereas at high k_{P-} it is the opposite one. Results in ref. [13] depict this type of diagram for dimensional variables with similar corresponding parameter values and only for the stable stationary states

solutions, over the range of values explored. The result is the response curve or bifurcation diagram (Fig. 3).

3. As observed when the nullclines for different parameter values are evaluated, modifications of values in different parameters can yield different response curves. With the knowledge gathered from the nullclines, repeat the previous **steps 1** and **2** for different parameter values. Note the different possible behaviors. For instance, decrease of d_C can change the bistable bifurcation diagram to an ultrasensitive curve. Compare Fig. 3 for $d_C = 1$ to Fig. 4 for $d_C = 0.05$. Another result obtained for other parameter values is shown in Fig. 5. When any model is analyzed, it is worth taking note of interesting or unexpectedly observed behaviors, because they can be converted into predictions and proposals of experiments.
4. Analyze variants of the model that could correspond to other available experimental data. For instance, compare the result when a protein is eliminated by setting its production to zero

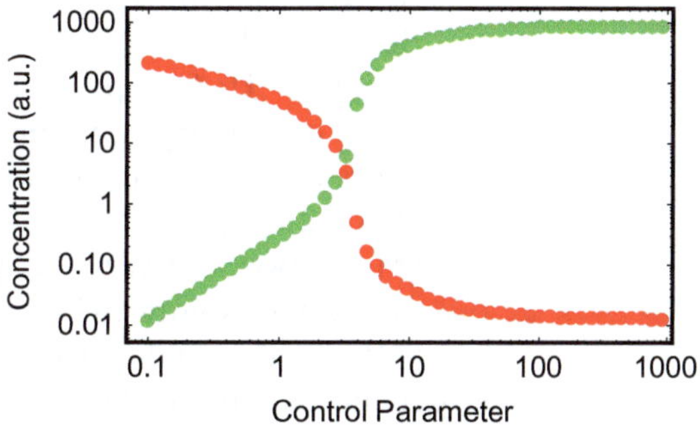

Fig. 4 Response curve/bifurcation diagram for active BES1 (*green*) and BRAVO (*red*) when the control parameter is the dephosphorylation rate k_{P-}, $d_C = 0.05$, and the remaining parameters are at their default value. The response is ultrasensitive

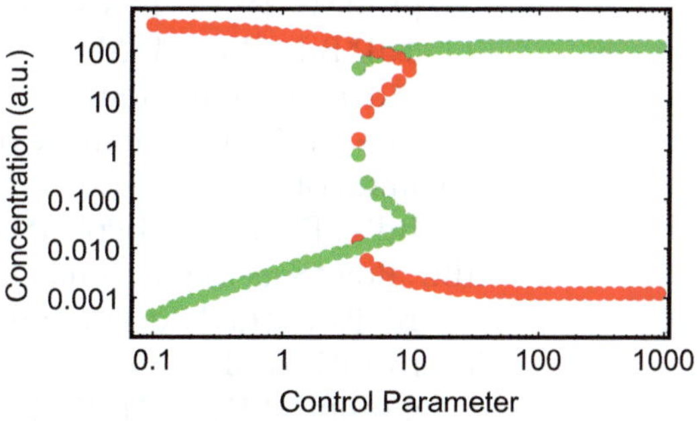

Fig. 5 Response curve/bifurcation diagram for active BES1 (*green*) and BRAVO (*red*) when the control parameter is the dephosphorylation rate k_{P-}, $k_{P+} = 10$, $d_{Bd} = 1$, and the remaining parameters are at their default value

or in some other way, which corresponds to modeling a mutant of this protein and can eventually enable comparison with experimentally studied mutants. In the BES1-BRAVO circuit, possible mutants are the *bes1-D* gain-of-function mutants that can be typified by increasing strongly the dephosphorylation rate k_{P-} and, correspondingly, keep BRAVO at a low level, as seen from the code results, or by the inclusion of a second BRAVO promoter obtained by doubling the BRAVO basal production α that makes the high BRAVO state more resilient to BES1 activation. When this type of analysis is done for any given model and the results are not in agreement with the experimental data, then one is forced to check the model assumptions and parameter choices and to build the model again, by repeating the whole method.

Mathematica code for computing the stationary solutions:

```
(***STATIONARY STATE CALCULATION***)
(*We initialize empty lists that will save the fixed points*)
Clear[M];Clear[Bd];Clear[B];Clear[Compl];
Bravo1 = {}; Bravo2 = {}; Bravo3={};
Bes1 = {}; Bes2 = {}; Bes3={};
Comp1 = {}; Comp2 = {}; Comp3={};
(*We can change parameters here if we want to see different
behaviors or go back to the first block and redefine them*)
(*dC=2.0;*)
(*In a loop, we vary a control parameter (param1) and com-
pute stationary solutions*)
(*If the loop takes too long to complete, you can put in a
larger increment, or decrease the exploration range by changing
the starting or ending points*)
For[param1=0.1 (*Starting point*),param1<1000(*Final poi
nt*),param1=param1*1.2(*increment*),
kPminus=param1;(*Now we choose that the control parame-
ter is k_{p-}*)
(*production term for M, i.e., BRAVO*)
prodM=α*(1+ε1*KM*M+ε2*KB*Bd+ε3*KM*M*KB*Bd)/
((1+KB*Bd) (1+KM*M));
(*dynamic equations*)
M'=prodM+kDminus*Compl-kDplus*M*Bd-M; (*Sets the
dynamics of BRAVO (Eq. 1)*)
Bd'=kPminus*B-kPplus*Bd+kDminus*Compl-kDplus*M*Bd-
dBd*Bd;(*Sets the dynamics of active BES1 (Eq. 2)*)
B'=β+kPplus*Bd-kPminus*B-dB*B; (*Sets the dynamics of
inactive BES1 (Eq. 3)*)
Compl'=kDplus*Bd*M-kDminus*Compl-dC*Compl;(*Sets
the dynamics of the BRAVO-BES1 Complex (Eq. 4)*)
(*Now find the stationary solutions, by setting derivatives to
zero and solving (restricting solutions to real, positive
numbers)*)
```

```
Sols=Solve[{M'==0,Bd'==0,B'==0,Compl'==0,M>=0,Bd>=0,
B>=0,Compl>=0},{M,Bd,B,Compl},Reals,WorkingPrecis
ion->20];
(*We save the value of the variables and control parameter to
lists*)
M1=M/.Sols[[1]];B1=B/.Sols[[1]];Bd1=Bd/.
Sols[[1]];C1=Compl/.Sols[[1]];
AppendTo[Bravo1,{param1,M1}];AppendTo[Bes1,{param1,
Bd1}];AppendTo[Comp1,{param1,C1}];
(*If there is more than one solution we also save the additional
solutions (up to three). Note that here we know a priori that there
will be a maximum of three solutions*)
If[Length[Sols]==3,M2=M/.Sols[[2]];
B2=B/.Sols[[2]];
Bd2=Bd/.Sols[[2]];
C2=Compl/.Sols[[2]];
M3=M/.Sols[[3]];
B3=B/.Sols[[3]];
Bd3=Bd/.Sols[[3]];
C3=Compl/.Sols[[3]];
AppendTo[Bravo2,{param1,M2}];AppendTo[Bes2,{param1,
Bd2}];AppendTo[Comp2,{param1,C2}];
AppendTo[Bravo3,{param1,M3}];AppendTo[Bes3,{param1,
Bd3}];AppendTo[Comp3,{param1,C3}]]]
(*We plot the solutions*)
If[Length[Bravo2]>0,
ListLogLogPlot[{Bravo1,Bes1,Bravo2,Bes2,Bravo3,Bes3
},PlotStyle->{Directive[AbsolutePointSize[8],Red],Directive[Abs
olutePointSize[8],Green]},Frame->True,FrameLabel->{"Control
Parameter","Concentration(a.u.)"},FrameStyle-
>Thick,LabelStyle->Directive[FontSize->18],ImageSize-
>400,PlotLegends->PointLegend[{"BRAVO","BES1"}]],ListLo
gLogPlot[{Bravo1,Bes1},PlotStyle->{Directive[AbsolutePointSiz
e[8],Red],Directive[AbsolutePointSize[8],Green]},Frame-
>True,FrameLabel->{"Control Parameter","Concentration
(a.u.)"},FrameStyle->Thick,LabelStyle->Directive[FontSize-
>18],ImageSize->400,PlotLegends->PointLegend[{"BRAVO","
BES1"}]]]
(*In the conditional we check whether the lists of the second and
third solutions are empty. If they are, we only plot the first one*)
```

3.4 Analyzing the Dynamic System

The next steps are done through the Mathematica code that is provided at the end of this subsection (*see* **Note 4**).

1. Evaluate the time evolution of the concentrations. Set initial values of the concentrations and run this part of the code (default values are provided in the code). The code performs the time integration of the dynamics by means of NDSolve and

by defining the dynamical equations with the explicit inclusion of time. When running the code, two figures appear: one that gives the concentration of each variable over time (Fig. 6) (*see* **Note 5**) and one that shows the trajectory of the values of BRAVO and BES1d, together with the nullclines (Fig. 7) (*see* **Note 6**). The trajectory ends at one stationary solution.

2. Try different initial conditions to see to which stationary solution the system evolves (*see* **Notes 7** and **8**).
3. Get a first glimpse on the stability of the stationary solutions found (Subheading 3.3) by setting several initial conditions around each stationary state. The solution is stable when all initial conditions around it end up at this solution.
4. Additional aspects can be evaluated with these tools when adapted to the model of interest. As examples, we name a few [1]. When interesting dynamical behaviors are observed, check the parameter ranges for which they happen and check whether

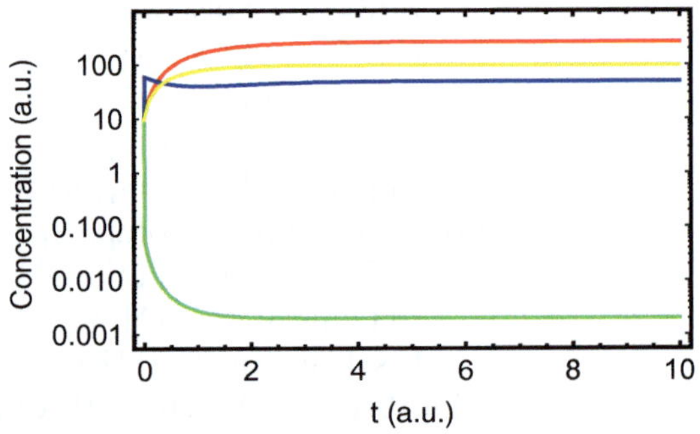

Fig. 6 Temporal evolution of the concentrations of active BES1 (*green*), BRAVO (*red*), inactive BES1 (*yellow*), and BRAVO-BES1 complex (*blue*) for the default values of initial conditions

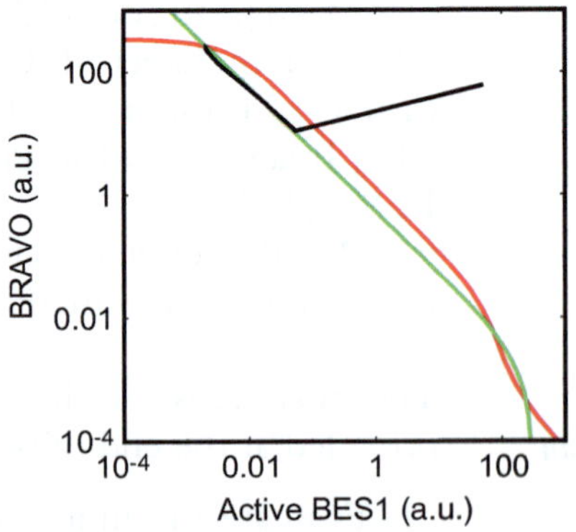

Fig. 7 Trajectory (*black*) of BRAVO and active BES1 values, corresponding to the temporal evolution depicted in Fig. 6, together with the nullclines (as in Fig. 1). The trajectory evolves toward a crossing point of the nullclines

they are consistent with experimental information on mutants, treatments, and other variants of the system [2]. If of interest to the particular system, analyze the response time scales of the system, namely which variable reaches the steady state faster and which more slowly; how does this depend on parameters; does it change in mutants [3]? The dynamical response of the system to a signal can also be studied: set a parameter that corresponds to the signal as a time-varying function, in whichever way that is suitable to represent the signal in the biological system, and evaluate the resulting dynamics.

The Mathematica code for obtaining the dynamics of the variables over time and their trajectory in the phase space is:

```
(***DYNAMICS***)
(*Clear the variables and derivatives*)
Clear[M];Clear[Bd];Clear[B];Clear[Compl];
M'=.;B'=.;Bd'=.;Compl'=.;Clear[t];
tmax=10;
(*The control parameter used before needs to be set at the desired value*)
kPminus=0.5;
(*dC = 0.02;
(*Parameters can be changed here. However, take care to use the same parameters as in the calculation of the nullclines if you want the last plot to make sense*)*)
(*To integrate dynamics we explicitly include time in the original equations, and put them together in a system of equations*)
(*dynamic equations, Eqs. 1–4*)
equations ={
D[M[t],t]==α*(1+ε1*KM*M[t]+ε2*KB*Bd[t]+ε3*KM*M[t]*KB*Bd[t])/((1+KB*Bd[t])*(1+KM*M[t]))+kDminus*Compl[t]-kDplus*M[t]*Bd[t]- M[t],
D[Bd[t],t]==kPminus*B[t]-kPplus*Bd[t]+kDminus*Compl[t]-kDplus*M[t]*Bd[t]-dBd*Bd[t],
D[B[t],t]==β+kPplus*Bd[t]-kPminus*B[t]-dB*B[t],
D[Compl[t],t]==kDplus*Bd[t]*M[t]-kDminus*Compl[t]-dC*Compl[t],
(*We also include in the equations the initial condition*)
M[0]== 60,Bd[0]== 50.0,B[0]== 10,Compl[0]== 10};(*here we can set the desired initial values*)
(*We can now use NDSolve to solve the differential equations*)
Vars ={M[t],Bd[t],B[t],Compl[t]};
Sol=NDSolve[equations,Vars,{t,0,tmax},MaxStepSize->0.001,MaxSteps->1000000];
```

```
Traj = LogPlot[{M[t]/.Sol,Bd[t]/.Sol,Compl[t]/.Sol,B[t]/.Sol},{t,0,tmax},PlotStyle->{Directive[Thick,Red],Directive[Thick,Green],Directive[Thick,Blue],Directive[Thick,Yellow]},Frame->True,FrameLabel->{"t (a.u.)","Concentration (a.u.)"},FrameStyle->Thick,LabelStyle->Directive[FontSize->18],ImageSize->400,PlotLegends->{"BRAVO","Active BES1","Complex","Inactive BES1"}]
(*Plot of the variables value over time*)
TrajMB = ParametricPlot[{Log[Bd[t]/.Sol[[1]]],Log[M[t]/.Sol[[1]]]},{t,0,tmax},PlotRange->LogRange,PlotStyle->Directive[Thick,Black],PlotLegends->{"Trajectory"}];
(*Plot of the trajectory in the space of M and Bd values*)
Show[NullclinesPlot,TrajMB,FrameLabel->{"Active BES1(a.u.)","BRAVO(a.u.)"}]
```

3.5 Proposed Experiments

According to the predictions made in Subheadings 3.3 and 3.4, experiments can be proposed that allow to discriminate between different assumptions of the model, i.e., that involve predictions that strongly change depending on these assumptions, are much more valuable, because they provide a disproportionate amount of information on the underlying working of the system. However, experiments that allow the observation of nontrivial behaviors of the system as predicted by the model are also useful, because they confirm that the model incorporates the essential elements and that the behaviors arising from these elements are informative about the biological system. In our example, we could propose different experiments according to the observed response of BES1 and BRAVO concentration to BES1 dephosphorylation. In particular, we predicted that BRAVO responded ultrasensitively to changes in the dephosphorylation rate, which are related to BR signaling [13]. We proposed an experiment that consisted in treating plants with different brassinolide (BL) concentrations and measured BRAVO expression as pBRAVO:GFP. BRAVO expression responded ultrasensitively to the amount of BL [13].

4 Notes

1. The number of parameters could be reduced by setting another appropriate value of C_0. Yet, we selected $C_0 = D_{tot1}$ to gain more intuitive comprehension of the transformation between dimensional and nondimensional variables. Additional reduction of parameters can be done by setting a different C_0 for each variable.
2. Parameter fitting to experimental data, when available, is desirable, for instance by minimizing the distance between the data and the model. This distance can be the sum of the square dif-

ferences divided by the error of each data point: (Datapoint value – Model value in same conditions)^2/(Datapoint Variance). For more details, see ref. [20]. In Eqs. 1–4, the nondimensional parameters are combinations of dimensional ones that can be estimated from empirical data (for estimation in this specific case, see ref. [13]) and of dimensional parameters of unknown value.

3. Nullclines are the curves that represent for which combined values of two variables x,y the time derivative of each of them is zero (there is a nullcline for each variable). Where the curves cross, both derivatives are zero, marking a stationary state.
4. Be aware of which parameter values were used last, because now these values will be utilized. We recommend to set the values to be used now in the initial part of the code (Subheading 3.1) and to run this initial part first, before running the additional code below.
5. For the set of default parameter values a very fast initial evolution is observed driven by the fast dimerization dynamics.
6. Be careful that this time evolution is computed for the same parameter values as those used to compute the nullclines, so that it makes sense to plot them all together.
7. The trajectory may depict some abrupt turns, because it only shows the trajectory in space of two variables, whereas the dynamics involve four variables.
8. Analysis of the trajectories to reach the steady states can be also of interest to observe in other dynamical models distinct from the one provided, namely behaviors, such as excitability and oscillations that may be nonintuitive, but linked to a biological function.

Acknowledgments

D.F. and M.I. acknowledge support from the Ministerio de Economía y Competitividad (Spain) and FEDER (EU) through grant FIS2015-66503-C3-3-P and from the Generalitat de Catalunya through Grup de Recerca Consolidat 2014 SGR 878. AIC-D acknowledges financial support from the Spanish Ministry of Economy and Competitiveness, through the 'Severo Ochoa Programme for Centres of Excellence in R&D' 2016–2019 (SEV-2015-0533). AIC-D is a recipient of a BIO2013-43873 grant from the Spanish Ministry of Economy and Competitiveness and European Research Council, ERC Consolidator Grant (ERC-2015-CoG – 683163).

References

1. Rué P, Garcia-Ojalvo J (2013) Modeling gene expression in time and space. Annu Rev Biophys 42:605–627
2. Kicheva A, Cohen M, Briscoe J (2012) Developmental pattern formation: insights from physics and biology. Science 338:210–212
3. Morelli LG, Uriu K, Ares S et al (2012) Computational approaches to developmental patterning. Science 336:187–191
4. Cruz-Ramírez A, Diaz-Triviño S, Blilou I et al (2012) A bistable circuit involving SCARECROW-RETINOBLASTOMA integrates cues to inform asymmetric stem cell division. Cell 150:1002–1015
5. Grieneisen VA, Scheres B, Hogeweg P et al (2012) Morphogengineering roots: comparing mechanisms of morphogen gradient formation. BMC Syst Biol 6:37
6. De Rybel B, Adibi M, Breda AS et al (2014) Integration of growth and patterning during vascular tissue formation in *Arabidopsis*. Science 345:1255215
7. van Esse GW, van Mourik S, Stigter H et al (2012) A mathematical model for BRASSINOSTEROID INSENSITIVE1-mediated signaling in root growth and hypocotyl elongation. Plant Physiol 160:523–532
8. Roeder AHK, Chickarmane V, Cunha A et al (2010) Variability in the control of cell division underlies sepal epidermal patterning in *Arabidopsis thaliana*. PLoS Biol 8:e1000367
9. Ellner SP, Guckenheimer J (2006) Dynamic models in biology. Princeton University Press, Princeton, NJ
10. Garcia-Ojalvo J (2011) Physical approaches to the dynamics of genetic circuits: a tutorial. Contemp Phys 52:439–464
11. Alon U (2006) An introduction to systems biology: design principles of biological circuits, Chapman & Hall/CRC Mathematical and Computational Biology, vol 10. Chapman & Hall/CRC Press, Boca Raton, FL
12. Strogatz SH (2014) Nonlinear dynamics and chaos: with applications to physics, biology, chemistry and engineering. Westview Press, Philadelphia, PA
13. Vilarrasa-Blasi J, Gonzalez-García MP, Frigola D et al (2014) Regulation of plant stem cell quiescence by a brassinosteroid signaling module. Dev Cell 30:36–47
14. Yin Y, Wang Z-Y, Mora-Garcia S et al (2002) BES1 accumulates in the nucleus in response to brassinosteroids to regulate gene expression and promote stem elongation. Cell 109:181–191
15. Vert G, Chory J (2006) Downstream nuclear events in brassinosteroid signalling. Nature 441:96–100
16. Becskei A, Séraphin B, Serrano L (2001) Positive feedback in eukaryotic gene networks: cell differentiation by graded to binary response conversion. EMBO J 20:2528–2535
17. François P, Hakim V (2004) Design of genetic networks with specified functions by evolution *in silico*. Proc Natl Acad Sci U S A 101:580–585
18. Buchler NE, Louis M (2008) Molecular titration and ultrasensitivity in regulatory networks. J Mol Biol 384:1106–1119
19. Wolfram Research, Inc. (2010) Mathematica, Version 8.0. Wolfram Research, Inc., Champaign, IL
20. Ashyraliyev M, Fomekong-Nanfack Y, Kaandorp JA et al (2009) Systems biology: parameter estimation for biochemical models. FEBS J 276:886–902

Chapter 10

Quantitative Microscopic Analysis of Plasma Membrane Receptor Dynamics in Living Plant Cells

Yu Luo and Eugenia Russinova

Abstract

Plasma membrane-localized receptors are essential for cellular communication and signal transduction. In *Arabidopsis thaliana*, BRASSINOSTEROID INSENSITIVE1 (BRI1) is one of the receptors that is activated by binding to its ligand, the brassinosteroid (BR) hormone, at the cell surface to regulate diverse plant developmental processes. The availability of BRI1 in the plasma membrane is related to its signaling output and is known to be controlled by the dynamic endomembrane trafficking. Advances in fluorescence labeling and confocal microscopy techniques enabled us to gain a better understanding of plasma membrane receptor dynamics in living cells. Here we describe different quantitative microscopy methods to monitor the relative steady-state levels of the BRI1 protein in the plasma membrane of root epidermal cells and its relative exocytosis and recycling rates. The methods can be applied also to analyze similar dynamics of other plasma membrane-localized receptors.

Key words Fluorescence, Confocal microscopy, FRAP, Brassinosteroids, BRI1, Plasma membrane, Exocytosis, Recycling

1 Introduction

Activation of mammalian plasma membrane-localized receptors, such as tyrosine kinases and G protein-coupled receptors, by binding their ligands at the cell surface induces endocytosis that can be a manner to decrease the number of receptors from the cell surface to attenuate signaling, to regulate the turnover of ligand-bound receptors from the plasma membrane, or to ensure signaling from the endosomes [1, 2]. Thus, ligand sensitivity and signaling strength are mostly regulated by ligand concentrations and plasma membrane receptor abundance.

In plant cells, the membrane-bound receptor BRASSINOSTEROID INSENSITIVE1 (BRI1) is a leucine-rich repeat (LRR) type of receptor kinase essential for plant growth, root development, reproduction, and cellular responses to biotic and abiotic stresses [3, 4]. BRI1 is activated by binding to the growth-promot-

Eugenia Russinova and Ana I. Caño-Delgado (eds.), *Brassinosteroids: Methods and Protocols*, Methods in Molecular Biology, vol. 1564,
DOI 10.1007/978-1-4939-6813-8_10,

ing plant steroid hormones, brassinosteroids (BRs), and its coreceptor BRI1-ASSOCIATED KINASE1 (BAK1) at the cell surface [5]. The ligand-bound BRI1 is internalized from the plasma membrane to the endosomes [6, 7]. However, unlike other plant LRR receptors of which the endocytosis depends on their ligands, such as FLAGELLIN-SENSITIVE2 [8], BRI1 undergoes constitutive (ligand-independent) endocytosis to the *trans*-Golgi network/early endosomes (TGN/EEs) and passes through the late endosomes/multivesicular bodies to accumulate into the vacuoles for degradation [6, 7, 9–11]. The half-life of the BRI1 receptor in root meristems is approximately 5–6 h [7, 12]. Mutation or overexpression analyses of the BRI1 protein or BR-biosynthetic enzymes support the concept that the output of the BRI1 signaling depends on the abundance of its ligand and the amount of plasma membrane-localized receptors [3, 13, 14]. Previously, we have used fluorescently labeled BR ligands to monitor BRI1 endocytosis in *Arabidopsis thaliana* root epidermal cells and we have shown that endocytosis inhibition (pharmacologically or genetically) of active BRI1 enhances the BR signaling from the plasma membrane [14, 15]. Like many cell surface-localized receptors, besides endocytosis and degradation, the plasma membrane pool of BRI1 is regulated by exocytosis and recycling. BRI1 constitutively cycles between the TGN/EEs and the plasma membrane and remains intracellular upon treatment with the fungal toxin brefeldin A (BFA) that interferes with exocytic protein sorting by reversible inhibition of the auxin response factor-guanidine nucleotide exchange factors (ARF-GEFs) [16] in BFA bodies [6, 7, 9, 11]. Upon washout of the drug, protein sorting of BRI1 to the plasma membrane resumes, as reflected by the gradual disappearance of the BFA-induced compartments [6, 7, 10, 11]. Thus far, BFA is the most commonly used pharmacological tool that interferes with recycling of plant plasma membrane proteins, including BRI1 [17]. BRI1 accumulation into BFA bodies, even in the presence of protein synthesis inhibitors [9, 18], is reversible upon washout, indicating that BRI1 continuously cycles between plasma membrane and recycling endosomes, of which the identity in plants is still unknown, but might partly constitute the TGN/EE. Consistently, disruption of BRI1 secretion and recycling by, for example, genetic and pharmacological interference with the pH of the TGN/EE reduced the BR signaling responses as a result of decreased BRI1 plasma membrane levels [18].

Here, we describe the procedures of quantitative microscopy methods to estimate the relative BRI1 plasma membrane levels in *Arabidopsis* root epidermal cells and to monitor BRI1 recycling. *Arabidopsis* plants expressing BRI1 fused to the green fluorescent protein (GFP) by its own promoter near endogenous expression levels were used to trace the BRI1 localization [7, 18]. To evaluate the contribution of recycling to the BRI1 plasma membrane pool, we measured the plasma membrane levels of BRI1-GFP in the presence of cycloheximide (CHX) and after a combined BFA and CHX treatment and calculated the recycling ratio as the relative reduction in plasma membrane fluorescence intensity after BFA application [18].

In contrast, we also report on the application of the quantitative fluorescence recovery after photobleaching (FRAP) method to measure BRI1 exocytosis in root epidermal cells [18]. FRAP is a powerful tool to study protein dynamics at the plasma membrane by measuring the fluorescence recovery of the bleached region of interest (ROI). In the plasma membrane, parts with the fluorescently tagged protein are bleached and their recovery is measured in the ROI. The recovery of the plasma membrane fluorescence in target cells depends on exocytosis, recycling, and lateral diffusion of nonbleached membrane components [19]. To avoid the impact of lateral diffusion and recycling insertion when analyzing BRI1 exocytosis, we selected a ROI that contained several cells and only measured cells in which the BRI1-GFP fluorescence had completely faded.

2 Materials

2.1 Plant Material

The *Arabidopsis thaliana* (L.) Heynh. (accession Columbia-0 [Col-0]) line expressing the *pBRI1::BRI1-GFP* construct [7].

2.2 Growth Media and Solutions

1. Liquid growth medium: half-strength Murashige and Skoog (½ MS) medium with 1 % (w/v) sucrose and 0.05 % (w/v) 2-(*N*-morpholino)ethanesulfonic acid (MES) monohydrate. For 1 L, weigh 2.2 g MS basal medium salts, 10 g sucrose, and 0.5 g MES. Adjust pH to 5.7 with KOH and add water to a final volume of 1 L. Autoclave to sterilize. Aliquot out liquid medium sterilely into 15-mL tubes for ease of use during treatments and microscopy.
2. Solid growth medium: ½ MS medium supplemented with 0.8 % (w/v) agar. Add 500 mL liquid medium to 500-mL glass bottles with 4 g agar. Autoclave to sterilize. Melt medium in a microwave and pour into square plates (50 mL per plate) as required.
3. 50 mM brefeldin A (BFA) stock solution prepared in dimethyl sulfoxide (DMSO) (*see* **Note 1**).
4. 50 mM cycloheximide (CHX) stock solution prepared in DMSO (*see* **Note 1**).
5. Commercial bleach.
6. Concentrated HCl.

2.3 Equipment

1. Confocal microscope equipped with 488-nm laser and ×60 objective (numerical aperture 1.2).

2. Growth chamber.
3. Fume hood.
4. 500-mL laboratory glass bottles.
5. Costar 12-well cell culture plates.
6. Imaging chamber (for instance, Nunc™ Lab-Tek™ II Chambered Borosilicate Coverglass).
7. Disposable plastic square plates, 100 × 100 mm.
8. Air-permeable tape width 2.5 cm.
9. Parafilm "M."
10. Microscope slides: 76 × 26 × 1 mm.
11. Cover slips: 22 × 50 mm.
12. Forceps.
13. 1.5-mL microcentrifuge tubes.

2.4 Analysis Software

1. Software from the confocal microscope manufacturer.
2. ImageJ (http://rsb.info.nih.gov/ij/).
3. GraphPad Prism (http://www.graphpad.com/scientific-software/prism/).

3 Methods

3.1 *Arabidopsis* Seed Sterilization and Plant Growth

1. Use the vapor-phase sterilization method to surface sterilize not more than 50 μL of *Arabidopsis* seeds in 1.5-mL microcentrifuge tubes.
2. Set an air-tight container in the fume hood and position the open seed-containing microcentrifuge tubes in the container.
3. Place a beaker with 100 mL commercial bleach inside the container.
4. Carefully add 3 mL concentrated HCl to the bleach and immediately seal the container.
5. Sterilize the seeds by chlorine fumes for 6–8 h (*see* **Notes 2** and **3**).
6. With a wet sterile toothpick, place seeds on the surface of the solid growth medium in square plates.
7. Seal the plates with air-permeable tape.
8. Store the plates with seeds for 3 days at 4 °C to synchronize germination.
9. Place the plates upright in a growth chamber at 22 °C with a 16-h/8-h light–dark photoperiod (110 μE/m^{2}s photosynthetically active radiation supplied by cool-white fluorescent tungsten tubes).

10. Let the seedlings grow for 4 days.

3.2 BRI1 Recycling Assay

3.2.1 Pharmacological Treatments

To evaluate the contribution of recycled BRI1 to the total plasma membrane receptor pool, we measure the plasma membrane levels of BRI1-GFP in the presence of CHX and after a combined treatment with BFA and CHX. The fungal toxin BFA and the protein synthesis inhibitor CHX are used to inhibit recycling/exocytosis in *Arabidopsis* cells [16] and to eliminate the newly synthesized BRI1 pool, respectively.

1. Prepare the 50 μM CHX working solution and the 50 μM CHX/50 μM BFA mix solution by diluting the relative stock solutions in ½ MS liquid medium.
2. Dilute the DMSO solution to 1 % (v/v) with the ½ MS liquid medium.
3. Use 12-well cell culture plates for the compound treatments.
4. Transfer 3 mL of the 1 % (v/v) DMSO solution to one well, 3 mL of 50 μM CHX to two wells, and 3 mL of the 50 μM CHX/50 μM BFA mix solution to one well.
5. Pipette 3 mL ½ MS liquid medium to each of the six wells.
6. Pick five seedlings from the ½ MS agar plate and place them into the well containing the 1 % (v/v) DMSO solution, taking care that all the roots are immersed.
7. Incubate the seedlings for 90 min at room temperature.
8. Wash seedlings thoroughly six times by transferring them from one well to another with ½ MS liquid medium.
9. Image seedlings immediately after wash.
10. Start the CHX treatment 30 min after the DMSO treatment to allow enough time for imaging of the DMSO control.
11. Pick five seedlings from the ½ MS agar plate and place them into one well containing 50 μM CHX, taking care that all the roots are immersed.
12. Incubate the seedlings for 90 min at room temperature.
13. Wash and image seedlings (*see* Subheading 3.2.1, **step 8**).
14. Start the CHX+BFA treatment 30 min after the CHX treatment.
15. Pick five seedlings from the ½ MS agar plate and place them into another well containing 50 μM CHX, taking care that all the roots are immersed.
16. Incubate the seedlings for 60 min at room temperature.
17. Transfer seedlings to the well containing the 50 μM CHX/50 μM BFA mix solution and incubate for 30 min at room temperature.

18. Wash and image seedlings (*see* Subheading 3.2.1, **step 8**).

3.2.2 Image Acquisition

1. Switch on the microscope with laser output for at least 10 min before use to eliminate power fluctuations during imaging.
2. With forceps carefully place the five seedlings on a clean microscope slide with 150–200 μL of liquid ½ MS medium so that the roots point in the same direction.
3. Gently place a glass cover slip, taking care to avoid air bubbles and squashing of the roots.
4. Wipe out excess medium.
5. Image on a laser scanning confocal microscope with a 60× objective.
6. Focus the sample in bright-field mode.
7. Set the imaging position consistently on the same area of the meristematic zone, 10–15 cells above the quiescent center (Fig. 1a) (*see* **Note 4**).
8. Focus on the root epidermal cell layer that is close to the objective lens (*see* **Note 4**).
9. Select enhanced (e)GFP from the dye list.
10. Set the 488-nm excitation laser power at 11 % and the emission channel at 500–550 nm (*see* **Note 5**).
11. Set the scan speed at 8.0 μs/pixel and the image size at 512 × 512 pixels.
12. Select the color mode for image saturation check (*see* **Note 6**).
13. Adjust the Offset and HV from the Image Acquisition Control panel to avoid the zero or saturated pixels (*see* **Note** 7).
14. Use the zoom in function to set the digital zoom at 3 and rotate the root to the vertical position (*see* **Note 8**).
15. Capture images and save in the default format that keeps all acquisition tags, including the confocal microscope settings.

3.2.3 Image Analysis and Quantifications

1. Use the confocal microscope software to export the TIFF files under saved image format.
2. Start the ImageJ program by opening the exported TIFF files.
3. Set image type to 8-bit (Image→Type→8-bit).
4. Set the measurement dialog to specify what will be recorded (Analysis→Set Measurements→select: "Area, Mean grey value, Mean and Max grey value") (*see* **Note 9**).
5. Use rectangle tool to select the ROIs for both the plasma membrane and the background (*see* **Note 10**).
6. Save both ROIs with the ROI manager (Analysis→Tools→ROI manager→Add) (*see* **Note 11**).

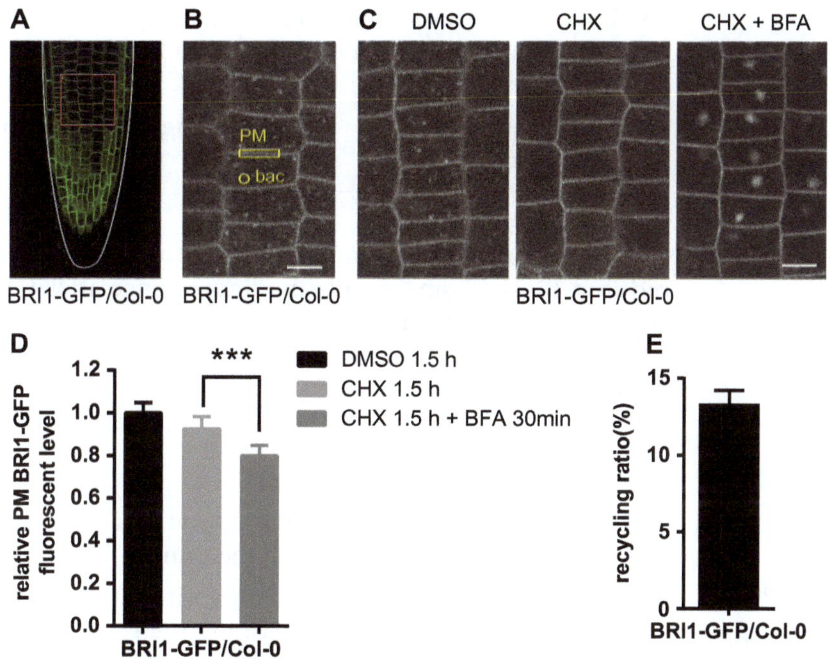

Fig. 1 Quantification of BRI1-GFP recycling ratio in root epidermal cells of *Arabidopsis*. (**a**) Imaging position is marked by the *red box*, 10–15 cells above the quiescent center. (**b**) *Gray-type* images of untreated BRI1-GFP/Col-0 root epidermal cells for fluorescence quantification. Plasma membrane ROI and background ROI are highlighted by *yellow frames* marked 'PM' and 'bac,' respectively. Scale bar, 5 μm. (**c**) BRI1-GFP fluorescence of BRI1-GFP/Col-0 treated with DMSO (1 %) for 90 min, CHX (50 μM) for 90 min, and BFA for 30 min in the presence of CHX for 90 min. Scale bar, 5 μm. (**d**) Graph representing the quantitative results of the relative plasma membrane fluorescence for BRI1-GFP/Col-0 after the different treatments presented in (**c**). ***$P < 0.001$. (**e**) Quantification of the recycling ratio of the BRI1-GFP protein in BRI1-GFP/Col-0. Error bars indicate S.D

7. Press "M" to get the measurement result (*see* **Note 12**).
8. Copy the data of the Mean values to an Excel sheet for further analysis.
9. Take at least one image from each of the five roots (i, i = 1 ~ 5).
10. For each image, calculate the average intensity of five plasma membrane ROIs (PM_i) and that of five background ROIs (bac_i).
11. Then calculate the average plasma membrane fluorescence intensity of five roots: ${}_{ave}PM = \left(\sum_{i=1}^{5} \frac{PM_i}{bak_i} \right) / 5$.
12. Calculate the average epidermal plasma membrane fluorescence of seedlings treated with DMSO (${}_{ave}PM^{DMSO}$), CHX (${}_{ave}PM^{CHX}$), and CHX+BFA (${}_{ave}PM^{CHX+BFA}$).
13. Copy the calculation results into GraphPad Prism.

14. Use the *t*-tests function (Analysis→Column analysis→t tests→unpaired) for the significant analysis of different treatments (*see* **Note 13**).
15. Calculate the recycling ratio as: $\frac{{}_{ave}PM^{CHX} - {}_{ave}PM^{CHX+BFA}}{{}_{ave}PM^{CHX}} \times 100\%$.

Examples of images and calculations are shown in Fig. 1. The reduced BRI1 plasma membrane fluorescence after CHX application when compared to the DMSO treatment is due to the inhibited de novo protein synthesis, whereas the decreased BRI1 plasma membrane fluorescence after the combined CHX+BFA treatment versus the CHX treatment alone reflects the recycled BRI1 to the plasma membrane.

3.3 Quantitative FRAP Analysis for BRI1 Exocytosis

3.3.1 Image Acquisition

1. Carefully place the seedlings in the imaging chamber with 150–200 μL of liquid growth medium.
2. Gently cover the whole seedlings with a piece of ½ MS agar, avoiding air bubbles and squashing the roots (*see* **Note 14**).
3. Image with laser scanning confocal microscope with a 60× objective (*see* Subheading 3.2.2).
4. Select eGFP from the dye list.
5. Set the 488-nm excitation laser power at 11 % and the emission channel at 500–550 nm (*see* **Note 14**).
6. Set the scan speed at 8.0 μs/pixel and the image size at 512 × 512 pixels.
7. Set the zoom value to 1.
8. Take a control image before bleaching.
9. Select a circular bleaching ROI of 35 μm in diameter.
10. Bleach with 100 % of the 488-nm line ray for 90 s (Fig. 2a).
11. Record the fluorescence recovery at 0, 10, 20, 30, 40, 50, and 60 min after bleaching with 11 % of the 488-nm laser.
12. Repeat the same bleaching and recovery imaging procedure for five independent seedlings.

3.3.2 Image Analysis and Quantifications

1. Take images from five roots (i, i = 1 ~ 5) for each time point (t, t = pre, 0, 10, 20, 30, 40, 50, and 60).
2. Measure plasma membrane fluorescence intensities of both totally bleached root epidermal cells (${}_{\text{bleach}}PM_i^t$) and cells outside the bleached ROI (${}_{\text{control}}PM_i^t$) for each root by ImageJ (*see* Subheading 3.2.3) (Fig. 2a).
3. Calculate the average plasma membrane fluorescence of bleached cells of five roots with: ${}_{ave\text{–bleach}}PM^t = \left(\sum_{i=1}^{5} \frac{{}_{\text{bleach}}PM_i^t}{{}_{\text{control}}PM_i^t} \right) / 5$.

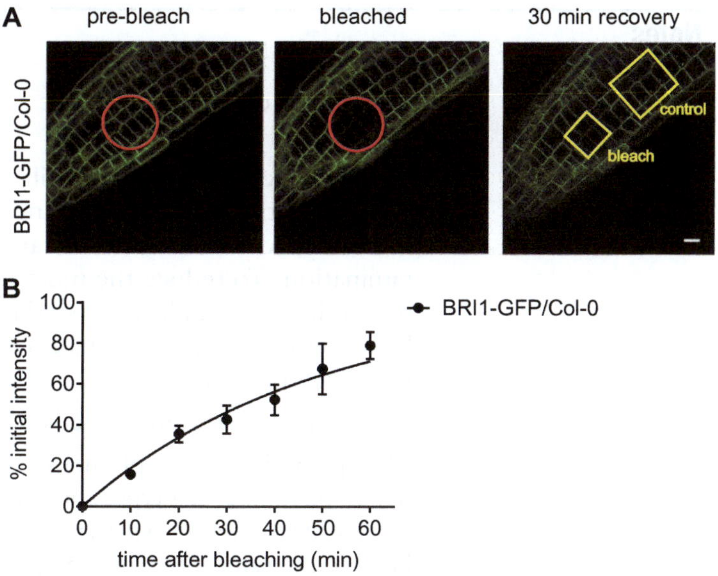

Fig. 2 FRAP of BRI1-GFP in root epidermal cells of *Arabidopsis*. (**a**) Images of BRI1-GFP/Col-0 before bleaching, immediately after bleaching, and after 30 min recovery of the GFP fluorescence after bleaching. The bleached areas are marked by *red oval frames*. The *yellow box* marked 'bleach' and the *box* marked 'control' highlights cells for bleached (${}_{bleach}PM_i^t$) and for control (${}_{control}PM_i^t$) plasma membrane measurement, respectively. Scale bar, 10 μm. (**b**) Quantitative FRAP analysis of BRI1-GFP/Col-0. Curve represents plasma membrane GFP fluorescence recovery rate after photobleaching of target cells. Error bars indicate S.D.

4. Set the relative plasma membrane fluorescence of bleached cells before bleaching (${}_{rel\text{-}bleach}PM^{pre}$) as 100 and immediately after bleaching (${}_{rel\text{-}bleach}PM^{0}$) as 0.
5. Calculate the relative plasma membrane fluorescence of bleached cells for each recovery time points (t = 10, 20, 30, 40, 50, 60) as:

$$ {}_{rel\text{-}bleach}PM^{t} = \frac{{}_{ave-bleach}PM^{i} - {}_{ave-bleach}PM^{0}}{{}_{ave-bleach}PM^{pre} - {}_{ave-bleach}PM^{0}} \times 100 . $$

6. Copy the calculation results to GraphPad Prism (Graph type: XY).
7. Plot the fluorescence recovery curves after bleaching by the nonlinear regression function (Analyze→XY analyses→ Nonlinear regression).

Examples of FRAP images and the recovery curves are shown in Fig. 2b.

4 Notes

1. The aliquoted stocks could be stored at −20 °C for up to 1 year.
2. The time period for seed sterilization is between 6 and 8 h; longer treatments with chlorine fumes would increase seed mortality, whereas shorter treatments might introduce contamination. To reduce the mortality and increase the shelf life of sterilized seeds, place the sterilized seed tubes in a clean flow bench with caps opened for 12 h to eliminate the chlorine remnants before storage.
3. The tubes with sterilized seeds can be capped and stored.
4. The quantification of plasma membrane fluorescence is sensitive to plant age, cell type, and growth conditions. To minimize these effects, we always carry out seed stratification, use seedlings of the same age, and image the same position of the root.
5. Adjust the gain and laser power from the confocal microscopy software to make sure the images are not overexposed. Image saturation would diminish fluorescence intensity differences between selected ROIs. Besides, high laser power could also cause fluorochrome-dependent cellular damage.
6. The procedure depends on the specific confocal microscope software. In the Live view, blue and red colors indicate zero pixel intensity and pixel saturation, respectively. The image acquisition requires no zero or saturated pixels.
7. Keep all the parameters constant for both confocal image acquisition (i.e., laser power, excitation wavelength, emission wavelength, photomultiplier voltage, gain, offset, pinhole size, scanning speed, image size, and digital magnification) and analysis (i.e., ROI size and shape).
8. The root cells need to be strictly vertical for the accurate plasma membrane fluorescence quantification. To ascertain that the root epidermal cells are vertical for the measurements, adjust the cells either by the confocal microscopy software when images are taken or by the image processing software (i.e., Microsoft Office Picture Manager) after imaging.
9. The average pixel intensity is displayed as Mean value in the "results" panel. The Min and Max value indicate minimum and maximum pixel intensity of the selected region. A Min value of 0 corresponds to underexposure and a Max value of 255 to overexposure.
10. The selected horizontal plasma membrane region should be thick enough to include the plasma membrane layer, but thin enough to exclude any endosomes. The selected background region within a cell should exclude any endosome (Fig. 1b).

11. Make sure to use the same plasma membrane and background ROIs for all image measurements.
12. To make sure the ROI remains the same among the measurements, just check the area size displayed in the "Results" window (Analysis → Set Measurements, → select: "Area, Mean grey value, Mean and Max grey value"). Once the area size is modified, go back to the ROI manager and double click the saved ROI to reload the original selection (Analysis → Tools → ROI manager).
13. $P < 0.05$ means statistically significant.
14. The imaging chamber and the agar cover provide seedlings with a comfortable environment for long-term imaging. For longterm observation of live samples in the FRAP assays, a constant and optimal growth environment (i.e., temperature and humidity) is important for the seedling conditions that could influence the endomembrane trafficking and the imaging results. It is advisable to use the imaging chamber and to cover the seedling with ½ MS agar pieces instead of glass slides to keep the seedlings under good conditions for extended periods.

Acknowledgements

We thank our colleagues Daniel Van Damme for help with the image analysis and critical reading of the manuscript, Evelien Mylle for technical assistance with the confocal microscope, and Martine De Cock for help in preparing the manuscript. Y.L. is indebted to the Belgian Science Policy Office (BELSPO) for a postdoctoral fellowship.

References

1. Sorkin A, von Zastrow M (2002) Signal transduction and endocytosis: close encounters of many kinds. Nat Rev Mol Cell Biol 3:600–614
2. Irannejad R, Tsvetanova NG, Lobingier BT et al (2015) Effects of endocytosis on receptor-mediated signaling. Curr Opin Cell Biol 35:137–143
3. Wang Z-Y, Seto H, Fujioka S et al (2001) BRI1 is a critical component of a plasma-membrane receptor for plant steroids. Nature 410:380–383
4. Vriet C, Russinova E, Reuzeau C (2012) Boosting crop yields with plant steroids. Plant Cell 24:842–857
5. Belkhadir Y, Jaillais Y (2015) The molecular circuitry of brassinosteroid signaling. New Phytol 206:522–540
6. Russinova E, Borst J-W, Kwaaitaal M et al (2004) Heterodimerization and endocytosis of Arabidopsis brassinosteroid receptors BRI1 and AtSERK3 (BAK1). Plant Cell 16:3216–3229
7. Geldner N, Hyman DL, Wang X et al (2007) Endosomal signaling of plant steroid receptor kinase BRI1. Genes Dev 21:1598–1602
8. Robatzek S, Chinchilla D, Boller T (2006) Ligand-induced endocytosis of the pattern recognition receptor FLS2 in *Arabidopsis*. Genes Dev 20:537–542

9. Dettmer J, Hong-Hermesdorf A, Stierhof Y-D et al (2006) Vacuolar H^+-ATPase activity is required for endocytic and secretory trafficking in *Arabidopsis*. Plant Cell 18:715–730
10. Viotti C, Bubeck J, Stierhof Y-D et al (2010) Endocytic and secretory traffic in *Arabidopsis* merge in the trans-Golgi network/early endosome, an independent and highly dynamic organelle. Plant Cell 22:1344–1357
11. Martins S, Dohmann EMN, Dompierre J et al (2015) Dual role for ubiquitin in plant steroid hormone receptor endocytosis. Nat Commun 6:6151
12. Geldner N (2011) Quantifying degradation rates of transmembrane receptor kinases. In: Dissmeyer N, Schnittger A (eds) Plant Kinases: Methods and Protocols, Methods in Molecular Biology, vol 779. Springer, New York, pp 217–224
13. Friedrichsen DM, Joazeiro CAP, Li J et al (2000) Brassinosteroid-insensitive-1 is a ubiquitously expressed leucine-rich repeat receptor serine/threonine kinase. Plant Physiol 123: 1247–1255
14. Irani NG, Di Rubbo S, Mylle E et al (2012) Fluorescent castasterone reveals BRI1 signaling from the plasma membrane. Nat Chem Biol 8:583–589
15. Di Rubbo S, Irani NG, Kim SY et al (2013) The clathrin adaptor complex AP-2 mediates endocytosis of BRASSINOSTEROID INSENSITIVE1 in *Arabidopsis*. Plant Cell 25:2986–2997
16. Geldner N, Friml J, Stierhof Y-D et al (2001) Auxin transport inhibitors block PIN1 cycling and vesicle trafficking. Nature 413:425–428
17. Luschnig C, Vert G (2014) The dynamics of plant plasma membrane proteins: PINs and beyond. Development 141:2924–2938
18. Luo Y, Scholl S, Doering A et al (2015) V-ATPase-activity in the TGN/EE is required for exocytosis and recycling in *Arabidopsis*. Nat Plants 1:15094
19. Hildick KL, González-González IM, Jaskolski F et al (2012) Lateral diffusion and exocytosis of membrane proteins in cultured neurons assessed using fluorescence recovery and fluorescence-loss photobleaching. J Vis Exp 60:e3747

Chapter 11

Analysis of In Vitro DNA Interactions of Brassinosteroid-Controlled Transcription Factors Using Electrophoretic Mobility Shift Assay

Simon J. Unterholzner, Wilfried Rozhon, and Brigitte Poppenberger

Abstract

Most signaling cascades ultimately lead to changes in gene expression by modulating the activity of transcription factors (TFs). The electrophoretic mobility shift assay (EMSA) is a simple but powerful in vitro method for investigation of specific protein–DNA interactions. It makes use of the fact that protein–DNA complexes have a lower electrophoretic mobility in gels than free DNA has. The application of labeled probes in combination with unlabeled competitors allows investigation of DNA-binding specificity and identification of binding motifs with single base-pair resolution. Here we describe the application of EMSAs for the study of interactions of the brassinosteroid-regulated TFs, BRASSINAZOLE-RESISTANT1, (BZR1), BRI1-ETHYL METHANESULFONATE-SUPPRESSOR1 (BES1)/BZR2, and CESTA with putative binding sites. The classical approach using radiolabeled probes, as well as the more recent application of fluorescent probes, is described and the advantages and disadvantages of both methods are discussed.

Key words Electrophoretic mobility shift assay, Probe, Competitor, Fluorescence, Protein expression, Transcription factor

1 Introduction

Brassinosteroids (BRs) regulate the expression of numerous genes; contribute to the regulation of plant development, cell elongation, and differentiation; impact on metabolic pathways; and participate in the control of stress responses [1]. To understand the versatile roles of the hormones, it is important to identify their direct targets. Genetic studies have revealed a significant role of the transcription factors (TFs) BRASSINAZOLE RESISTANT1 (BZR1), BRI1-ETHYL METHANESULFONATE-SUPPRESSOR1/ BRASSINAZOLE RESISTANT2 (BES1/BZR2), and CESTA (CES) in the regulation of BR-responsive gene expression [2–8]. A central approach that has been used to elucidate the DNA-binding activities of these TFs is the electrophoretic mobility shift assay

Eugenia Russinova and Ana I. Caño-Delgado (eds.), *Brassinosteroids: Methods and Protocols*, Methods in Molecular Biology, vol. 1564, DOI 10.1007/978-1-4939-6813-8_11, © Springer Science+Business Media LLC 2017

(EMSA). EMSA is a powerful and sensitive method for in vitro verification of TF–DNA interactions, which, in addition, can also be used for the identification of binding sites at single base-pair resolution [9].

The assay is based on the observation that the electrophoretic mobility of DNA–protein complexes is slower than the mobility of free nucleic acids. In order to study such interactions, the labeled DNA probe and the protein of interest are mixed and incubated under conditions allowing binding. Subsequently, they are electrophoresed using a polyacrylamide gel. As a control, a labeled probe without protein is loaded on the same gel. Comparison of the electrophoretic mobility of the labeled nucleic acid probe in the absence and presence of the protein allows the determination of whether the probe and the protein interact. In case of binding, an additional band of lower mobility representing the DNA–protein complex appears. Moreover, the specificity of interaction can be easily studied by adding unlabeled competitor DNA fragments, usually synthetic oligonucleotides, to the binding reaction (*see* Fig. 1). Addition of oligonucleotides containing the binding site in 10- to 100-fold molecular excess will diminish or completely block the band shift, whereas addition of oligonucleotides containing mutated binding sites should have no or only a negligible impact on the band shift. By using a series of oligonucleotides containing different mutations, it is also possible to characterize the binding site of the protein [10].

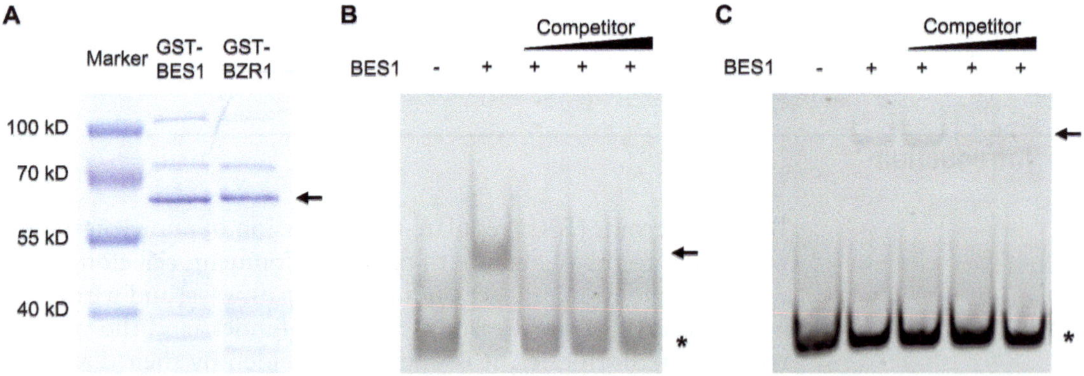

Fig. 1 Binding of BES1 to the promoter of *GA20ox1*. (**a**) SDS-PAGE of purified recombinant GST-BES1 and GST-BZR1 fusion proteins purified from *E. coli* BL21. The expected size of GST-BES1 and GST-BZR1 is 63.5 kDa and is marked by the *arrow*. (**b**) Binding of GST-BES1 to a 101-bp radiolabeled fragment of the *GA20ox1* promoter. The free probe is marked with an *asterisk* and the GST-BES1-DNA complex with an *arrow*. Addition of a 10, 100, and 1000-fold molar excess of a 36-bp competitor containing the BES1/BZR1-binding site suppresses the shift. Please mind that the diffuse band between the probe and the probe–DNA complex is not a shift, but an impurity already present in the control without protein. (**c**) Fluorescence-labeled EMSA with GST-BES1 using a HEX-labeled 151-bp fragment of the *GA20ox1* promoter as probe. Addition of the same competitor as in (**b**) at a 1-, 10-, or 100-fold molar excess diminishes and, eventually, suppresses the shift. *Symbols* have the same meaning like in (**b**)

Whereas EMSA is mainly used for qualitative studies of protein–DNA complexes, it also allows the measurement of the binding stoichiometry, affinity, and kinetics [11–13]. EMSA is very sensitive and can be performed with small amounts of protein and DNA probes, making it very suitable for verification of direct interactions of TFs to putative DNA binding sites. A major limitation of the method is that the protein–DNA complex needs to be stable during the electrophoresis step, but at low temperatures most TF–DNA complexes are sufficiently stable.

Traditionally, DNA probes were labeled with radioisotopes, particularly ^{32}P or ^{33}P. Radiolabeling offers a very high sensitivity and a simple detection by autoradiography. However, due to safety considerations, high cost for radionucleotides, and the very short lifetime of radiolabeled probes, a number of alternatives have been developed, including chemiluminescence [14, 15], immunolabeling [16], and fluorescence labeling [17, 18]. Due to the availability of fluorescence-labeled oligonucleotides, the latter technique is particularly simple, cost efficient, and has a good sensitivity. In addition, the labeled probes can be stored almost indefinitely, making this approach very convenient for occasionally performed experiments.

Here protocols for EMSA with radio- and fluorescence-labeled DNA probes are provided. In both cases, 5′-labeled probes are used: a radioactive phosphate is attached to the dephosphorylated probe with polynucleotide kinase (PNK) and [γ-^{33}P]-ATP, whereas fluorescent probes are generated by polymerase chain reaction (PCR) with 5′-hexachlorofluorescein (HEX)-labeled primers. HEX has an excitation maximum of 535 nm and an emission maximum at 556 nm, allowing detection with a 532-nm diode laser for excitation and a suitable long-pass filter, usually 550 nm, for selection of emitted light. Both methods have been applied to study binding of BES1, BZR1, and CES to promoters of BR-responsive genes in vitro [4, 10].

2 Materials

2.1 Protein Purification Components

1. Modified Luria broth (MLB) medium: 10 g/L peptone, 5 g/L yeast extract, 5 g/L NaCl, pH 7.0–7.5. Autoclave at 121 °C for 20 min.
2. 10× phosphate buffered saline (PBS): 80 g/L NaCl, 2 g/L KCl, 14.4 g/L Na_2HPO_4, 2.4 g/L KH_2PO_4; pH 7.4.
3. Lysozyme solution: 100 mg/mL dissolved in water. The solution should be prepared freshly.
4. Isopropyl-β-d-thiogalactopyranoside (IPTG) solution: 1 M IPTG dissolved in water. Store at −20 °C.

5. Phenylmethylsulfonyl fluoride (PMSF) solution: 400 mM PMSF dissolved in dimethylsulfoxide (DMSO). The solution may be stored at −20 °C for up to 6 months.
6. 1,4-dithiothreitol (DTT) solution: 1 M DTT dissolved in water. Store at −20 °C.
7. Triton X-100 solution: 20 % (v/v) dissolved in water. Store at 4 °C.
8. Elution buffer: 150 mM NaCl, 5 mM DTT, 20 mM glutathione reduced form (GSH), 50 mM TRIS–HCl, pH 8.0; pH must be checked and, if necessary, set to pH 7.5–8.0 with 1 M NaOH. The solution can be stored at −20 °C for up to 6 months.
9. Ampicillin solution: dissolve 1 g ampicillin sodium salt to 10 mL in water and sterilize by filtration. Store at −20 °C for up to 6 months.

2.2 Probe Preparation and Labeling Components

1. T4 polynucleotide kinase (PNK) 10 U/μL and 10× PNK buffer for forward reaction.
2. [γ-^{33}P]-ATP, 3000 Ci/mmol, 10 mCi/mL.
3. 10× EMSA binding buffer: 250 mM N-(2-hydroxyethyl) piperazine-N′-(2-ethanesulfonic acid) (HEPES) set with KOH to pH 8.0, 500 mM KCl, 20 mM $MgSO_4$, and 10 mM DTT.
4. 50 % (v/v) glycerol.
5. Developer: Kodak GBX developer/replenisher concentrate, 1 to 10 diluted with water.
6. Fixer: Kodak GBX fixer concentrate, 1 to 10 diluted with water.

2.3 DNA Quantification Components

1. DNA standard (10 ng/μL): mix 33.3 μL λ of DNA (0.3 μg/μL with 966.7 μL water. Keep at −20 °C.
2. TRIE buffer: 25 mM triethanolamine, 5 mM ethylenediaminetetraacetic acid (EDTA)-free acid.
3. SYBR stock solution: dilute 10 μL SYBER green I nucleic acid gel stain with 990 μL DMSO. The solution can be stored at −20 °C for at least 2 years.
4. SYBR reagent: add 20 μL SYBR stock solution to 20 mL 1× TRIE buffer. The solution must be prepared freshly.

2.4 Gel Electrophoresis Components

1. 10× TBE Stock: 108 g/L TRIS base, 55 g/L boric acid, 9.3 g/L EDTA, pH 8.3.
2. 10 % (w/v) ammonium peroxodisulfate (APS): dissolve 1 g APS in water to a total volume of 10 mL. The solution can be stored at 4 °C for up to 1 month.
3. *N*,*N*,*N*′,*N*′-tetramethylenediamine (TEMED).

4. Acrylamide stock solution: dissolve 38.0 g acrylamide and 1.04 g *N,N'*-methylenebis(acrylamide) in water to a total volume of 100 mL and filter through a 0.45-μm filter. Alternatively, a commercial 40 % (w/v) acrylamide solution with a ratio of acrylamide to *N,N'*-methylenebis(acrylamide) of 37.5:1 can be used.
5. 5× DNA loading buffer: 60 % (v/v) glycerol, 0.05 % (w/v) bromophenol blue sodium salt.

2.5 Other Equipment and Materials

1. Thermocycler (such as Mastercycler®, Eppendorf).
2. Molecular Imager (such as FX Pro, Biorad) equipped with a 532-nm laser and a 550-nm long-pass (LP) filter.
3. 3MM paper (Whatman).
4. Vacuum gel drier.
5. X-ray film.
6. Gel electrophoresis apparatus for vertical glass plates 20 × 20 cm, cooled, and with 1-mm spacers and combs (such as Vertical Double Gel Electrophoresis Twin L, PEQLAB Biotechnologie GmbH).
7. Fluorimeter.
8. Microcuvette: 1 cm width, 1 mL volume.
9. Refrigerated table top centrifuge.
10. Refrigerated microcentrifuge.
11. Sonicator for cell lysis.
12. Glutathione beads.
13. Spin filter column (such as Roti-spin MINI-3).
14. Centrifugal devices for ultrafiltration for a sample volume of at least 500 μL and a molecular mass cut-off of 3 kDa.
15. Silica membrane-based gel extraction kit.

3 Methods

3.1 Purification of BES1, BZR1, and CES as GST Fusion Proteins

The method described later was tested for purification of BES1, BZR1, and CES as GST-fusion proteins. The coding sequences of these genes were cloned into pGEX vectors that were subsequently transformed into *Escherichia coli* BL21 cells [4, 19]. An example for the purification of BES1 and BZR1 as GST fusion proteins is shown in Fig. 1a.

3.1.1 Induction of Protein Expression

1. Inoculate approximately 20 mL of 100 mg/L ampicillin-supplemented MLB with a single colony of *E. coli* BL21 containing the desired expression construct. Incubate the culture on a shaker at 22 °C overnight.

2. Inoculate 200 mL MLB with the overnight culture to an optical density of 0.2 (OD_{600}) and grow at 22 °C to an OD_{600} of 0.5.
3. Add IPTG to the culture to a final concentration of 0.1 mM.
4. Incubate the culture at 22 °C for 5 h, harvest the cells by centrifugation at 3000 × *g* for 10 min and freeze the pellet at −20 °C.

3.1.2 Protein Purification

All steps have to be performed at 4 °C or on ice.

1. Add 10 mL ice-cold 1× PBS containing 2 mM PMSF and 5 mM DTT to the pellet and shake until no cell clumps remain.
2. Add 200 μL of a freshly prepared solution of lysozyme (100 mg/mL in water) and incubate on ice for 10 min.
3. Add 500 μL 20 % (v/v) Triton x-X00 and mix well by inverting the tube several times before sonication of the cells four times for 10 s each and an output power of 35 % (*see* **Note 1**).
4. Centrifuge the sonicated cells at 20,000 × *g* for 10 min and transfer the supernatant to a new tube and centrifuge again (20,000 × *g* for 10 min).
5. In the meantime, wash 200 μL GSH beads (*see* **Note 2**) in a 15-mL screw cap tube with 10 mL PBS containing 5 mM DTT. Centrifuge at 200 × *g* for 1 min and discard the supernatant.
6. Transfer the clear supernatant of the cell lysate to the washed GSH beads and incubate on a shaker at 4 °C for 1 h.
7. Pellet the GSH beads by centrifugation at 200 × *g* for 1 min and remove the supernatant. Wash the GSH beads four times with 10 mL 1× PBS containing 5 mM DTT and incubate at 4 °C on a shaker for 5 min between the washes.
8. Resuspend the washed beads in approximately 1.5 mL of 1× PBS containing 5 mM DTT and transfer them completely to a 2-mL reaction tube, pellet the GSH beads by centrifugation at 200 × *g* for 1 min and remove the supernatant as completely as possible.
9. Add 500 μL elution buffer and incubate at 4 °C on a shaker for 30 min. Pellet the GSH beads by centrifugation at 200 × *g* for 1 min and transfer the clear supernatant to a 2-mL reaction tube. Repeat this step once.
10. Transfer the clear supernatant to a spin filter column and centrifuge at 200 × *g* for 2 min.
11. Load the flow-through onto a spin filter column and centrifuge for 15 min at 16,000 × *g* until 80–100 μL liquid is left. Aliquot the liquid in volumes of 10 μL each into small tubes and store at −80 °C. Estimate the protein concentration by

sodium dodecyl sulfate-polyacrylamide gel electrophoresis (SDS-PAGE) using bovine serum albumin (BSA) as a standard.

3.2 EMSA with Radiolabeled Nucleic Acid Probes

This method was used for investigation of the binding activities of BES1 and BZR1 to the gibberellin 20 oxidase 1 (*GA20ox1*) promoter [10]. BES1 in complex with a radiolabeled fragment of the *GA20ox1* promoter is shown in Fig. 1b. Addition of mutated competitor fragments (as in Fig. 1b) allowed for the detailed characterization of the binding motif [10].

3.2.1 Radioactive Labeling of the EMSA Probe

1. Amplify the region to be investigated by PCR (*see* **Notes 3** and **4**).
2. Load the amplicon on an agarose gel, electrophorese, and purify the PCR product using a silica membrane-based gel extraction kit (*see* **Note 5**).
3. Use approximately 1 ng of the purified DNA as a template for a second round of PCR using the same primers as before (*see* **Note 6**). If required, prepare several PCR reactions in parallel, because several hundred ng amplicons are required.
4. Purify the amplicon using a silica membrane-based gel extraction kit (*see* **Note** 7). The amplicon can be stored at −20 °C indefinitely.
5. Measure the DNA concentration by fluorimetry (*see* Subheading 3.2.2).
6. Transfer 2 pmol DNA of the purified amplicon (*see* **Note 8**) into a 1.5-mL tube, add 2 μL 10× PNK forward buffer (*see* **Note 9**), 5 pmol [γ-^{33}P]-ATP (*see* **Notes 10** and **11**), and water to a final volume of 19 μL. Subsequently, add 1 μL PNK, mix gently, and incubate at 37 °C for 30 min.
7. Purify the DNA probe with a silica membrane-based PCR product purification kit (*see* **Note** 7) and measure the DNA concentration (*see* Subheading 3.1.2). To assess the successful labeling, the activity may be measured by scintillation counting. Ideally, the activity should exceed 5000 cpm/pmol. The probe may be stored at −20 °C for a few days, but immediate use is recommended.

3.2.2 Quantification of DNA by Fluorimetry with SYBR-Green

Quantification of DNA by fluorimetry is recommended, because this technique is more sensitive than UV spectroscopy and the HEX fluorescence label (*see* Subheading 3.3) does not interfere.

1. Prepare for each standard and sample one plastic microcuvette filled with 1 mL SYBR reagent.
2. Add 0, 2, 4, 6, 8, and 10 μL of 10 ng/μL DNA standard and 1 μL (or another suitable volume) of the sample into cuvettes and mix properly.

3. Set the excitation wavelength of the fluorimeter at 492 nm and the emission wavelength at 519 nm and measure the fluorescence of the standards and sample to calculate the DNA concentration in ng/μL.

3.2.3 Protein–DNA Binding Reaction

1. Transfer 0.1 pmol radiolabeled DNA probe, 2 μL of 10× EMSA binding buffer, 2 μL of 50 % (v/v) glycerol, and, if desired, 2 μL of competitor oligonucleotides (*see* **Note 12**) to a 1.5-mL reaction tube and place on ice.
2. Add 200 ng of purified recombinant GST-fusion protein (*see* **Note 13**) and add water to a final volume of 20 μL. Incubate the mixture for 30 min on ice.

3.2.4 Gel Electrophoresis

1. Prepare a 6 % (w/v) acrylamide gel by mixing 7.5 mL 40 % (w/v) acrylamide (37.5:1), 5 mL 10× TBE buffer, 37 mL water, and 50 μL TEMED (the indicated volumes are sufficient for casting one 20 × 20 cm gel with 1-mm spacers).
2. Add 500 μL 10 % (w/v) APS and immediately cast the gel avoiding any air bubbles. Let polymerize for approximately 20 min.
3. Prerun the gel at 200 V and 50 mA for at least 1 h in an electrophoresis apparatus cooled below 4 °C (*see* **Note 14**). If more gels are run at the same time, the current must be increased accordingly, while the voltage is kept at 200 V.
4. Add 5 μL of 4× DNA loading buffer to the reaction, mix well, and load the reactions on the gel.
5. Apply electricity as described earlier (200 V, 50 mA per gel, below 4 °C) and let run until the bromophenol blue has migrated through approximately 2/3 of the gel (3–4 h).

3.2.5 Detection

1. Open the plate assembly and gently separate the two plates. Place a piece of 3MM paper on the gel, press softly, and remove the paper with the gel sticking to it.
2. Dry the gel mounted on 3MM paper in a vacuum drier.
3. Place the dried gel in an autoradiography cassette and expose an X-ray film to the gel. The exposure time depends on the activity and may vary from several hours to a few weeks (*see* **Note 15**).
4. Open the cassette and develop the film by placing it in diluted developer until the autoradiography signal is visible. Subsequently, wash the film with water and incubate it in diluted fixing solution for at least 10 min. Finally, place the film for at least 10 min in water before drying.

3.3 EMSA with Fluorescence-Labeled Nucleic Acid Probes

Fig. 1c shows the outcome of a BES1-*GA20ox1* promoter-binding experiment when, instead of a radiolabeled fragment (Fig. 1b), a fluorescence-labeled DNA fragment is used with the method described later.

3.3.1 Preparation of Fluorescence-Labeled EMSA Probes

1. Use HEX-labeled (*see* **Note 16**) primers (labeled forward and/or reverse primers can be used) for amplification of the region of interest by PCR. Several PCR reactions may be necessary to obtain the required amount of amplicon.
2. Load the amplicon on an agarose gel, electrophorese, and purify using a silica membrane-based gel extraction kit. If only the desired band is obtained, the amplicon may also be purified with a silica-membrane based PCR purification kit (*see* **Note 17**)
3. Measure the DNA concentration by fluorimetry (*see* Subheading 3.2.2). The final DNA concentration should be at least 10 ng/μL.
4. The probe can be stored in the dark at −20 °C for at least 2 years.

3.3.2 Protein–DNA Binding Reaction

1. Transfer 0.5 pmol HEX-labeled DNA probe, 2 μL of 10× EMSA binding buffer, 2 μL of 50 % (v/v) glycerol and, if desired, 2 μL of competitor DNA (*see* **Note 12**) into a 1.5-mL reaction tube and place on ice.
2. Add 200 ng of purified recombinant GST-fusion protein (*see* **Note 13**) and water to a final volume of 20 μL. Incubate the mixture for 30 min on ice.

3.3.3 Gel Electrophoresis

1. Gel electrophoresis is performed as described in Subheading 3.1, **step 4**.

3.3.4 Detection

1. Remove the gel sandwiched between the glass plates from the electrophoresis apparatus.
2. Wash the assembly carefully with water prior to drying its surfaces with a black cloth towel.
3. Place the gel sandwiched between the two glass plates in a molecular imager scanner and detect the fluorescence signal using the 532-nm laser and the 550-nm LP filter.

4 Notes

1. We use a Sonoplus HD2070 equipped with a UW 2070 probe (Bandelin). For other instruments, optimal settings may be different. Please refer to the recommendations of the manufacturer.

2. The type of GSH beads used may impact significantly on the yield of recombinant proteins. We obtained good results with GSH Sepharose 4B beads (GE Healthcare) and low yields with GSH cellulose beads.
3. Fragments with sizes of approximately 30–500 bp can be used as probes. However, binding may be less efficient for small fragments, whereas the shift may be small for large probes. Thus, we recommend probes with a size of approximately 100 bp.
4. Probes may also be generated by restriction digest of cloned fragments (in this case dephosphorylation with alkaline phosphatase is required) or by using synthetic DNA.
5. All kits we tested so far were suitable and yielded similar results, namely, E.Z.N.A.® MicroElute Gel Extraction Kit (Omega Biotek), QIAquick Gel Extraction Kit (Qiagen), and Wizard® SV Gel and PCR Clean-Up System (Promega).
6. This two-step PCR protocol is recommended because it removes template DNA and helps preventing formation of unspecific amplicons that both would be labeled by PNK and may give signals that might be confused with band shifts.
7. All kits we tested so far were suitable and yielded similar results, namely, E.Z.N.A.® Cycle Pure Kit (Omega Biotek), QIAquick PCR Purification Kit (Qiagen), and Wizard® SV Gel and PCR Clean-Up System (Promega)
8. The concentration of DNA can be calculated by the following formula: (ng DNA/bp DNA fragment) × 1.64 = pmol probe.
9. Two buffers are available for T4 PNK: buffer A and buffer B for forward and exchange reactions, respectively (Thermo Fisher Scientific Kit). Be sure to use the buffer for forward reaction, because it gives better labeling results for dephosphorylated fragments.
10. The same protocol can also be applied for labeling of probes with [γ-^{32}P]-ATP instead of [γ-^{33}P]-ATP.
11. With an activity of 3000 Ci/mmol and a concentration of 10 mCi/mL, the concentration of [γ-^{33}P]-ATP is 3.3 pmol/μL. In this case, 1.5 μL of [γ-^{33}P]-ATP should be used for labeling of 2 pmol probe. Increasing the amount of [γ-^{33}P]-ATP may enhance the labeling efficiency slightly.
12. For preparation of competitors, short complementary oligonucleotides (30–40 bp are usually sufficient) are ordered and hybridized by mixing equal volumes of the two 100 μM stock solutions. Thus, the final solution is 50 μM (= 50 pmol/μL) for the double-stranded competitor. The mixture is heated for 2 min to 94 °C, then cooled rapidly to 80 °C, and finally cooled at a rate of approximately −2 °C/min to room temperature.

Subsequently, the following dilutions are prepared for radioactive EMSAs: 2 μL hybridized oligonucleotides + 198 μL water, corresponding to 1 pmol/2 μL (10× excess) and 20 μL hybridized oligonucleotides + 190 μL water, resulting in a solution containing 10 pmol/2 μL (100× excess). For fluorescence-labeled EMSAs, dilutions of 5 μL double-stranded oligonucleotides + 95 μL water and 50 μL double-stranded oligonucleotides and 50 μL water, corresponding to 5 pmol/2 μL (10× excess) and 50 pmol/2 μL (100× excess) are prepared, respectively. Hybridized competitors and their dilutions can be stored at −20 °C indefinitely. For initial testing of whether binding happens, it is also possible to work without competitors.

13. The optimal concentration depends on the protein, its tag, and successful purification. Thus, it is useful to determine the best amount by titrating a constant amount of probe with different amounts of protein.
14. Many protein–DNA complexes, including that of BES1, BZR1, and CES are too labile at room temperature. Thus, cooling of the electrophoresis apparatus below 4 °C is required. In addition, the prerun removes the remaining APS from the gel.
15. For well-labeled probes, bands should be visible after 48 h of exposure or less. The significance of very weak bands that require several weeks of exposure has to be seen as very critical, because they may represent artifacts, for instance unspecific PCR bands or templates.
16. Instead of HEX-labeled primers, also Cy3-labeled primers may be used. However, in our hands, detection of HEX-labeled probes was more sensitive. HEX- or Cy3-labeled primers are designed by the same rules as unlabeled primers, but we recommend a minimal length of 24 nucleotides.
17. In contrast to radiolabeling, only the amplicons are labeled by this procedure. Thus, it is not necessary to take care that the template (cloned or genomic DNA) is removed. However, when unspecific PCR products appear, a two-step PCR protocol as described in Subheading 3.2.1 may be helpful. The same kits as mentioned in **Notes 3** and **5** can be used for purification of the amplicons.

Acknowledgements

The Deutsche Forschungsgemeinschaft (SFB924 TP A12 to B.P.) supported this work. S.J.U. was a member of the TUM graduate school.

References

1. Clouse SD (2011) Brassinosteroids. The Arabidopsis Book 9:e0151
2. Wang Z-Y, Nakano T, Gendron J et al (2002) Nuclear-localized BZR1 mediates brassinosteroid-induced growth and feedback suppression of brassinosteroid biosynthesis. Dev Cell 2:505–513
3. Yin Y, Wang Z-Y, Mora-Garcia S et al (2002) BES1 accumulates in the nucleus in response to brassinosteroids to regulate gene expression and promote stem elongation. Cell 109:181–191
4. Poppenberger B, Rozhon W, Khan M et al (2011) CESTA, a positive regulator of brassinosteroid biosynthesis. EMBO J 30:1149–1161
5. Yin Y, Vafeados D, Tao Y et al (2005) A new class of transcription factors mediates brassinosteroid-regulated gene expression in *Arabidopsis*. Cell 120:249–259
6. He J-X, Gendron JM, Sun Y et al (2005) BZR1 is a transcriptional repressor with dual roles in brassinosteroid homeostasis and growth responses. Science 307:1634–1638 [Erratum Science 308:1743]
7. Nemhauser JL, Hong F, Chory J (2006) Different plant hormones regulate similar processes through largely nonoverlapping transcriptional responses. Cell 126:467–475
8. Sun Y, Fan X-Y, Cao D-M et al (2010) Integration of brassinosteroid signal transduction with the transcription network for plant growth regulation in *Arabidopsis*. Dev Cell 19:765–777
9. Hellman LM, Fried MG (2007) Electrophoretic mobility shift assay (EMSA) for detecting protein-nucleic acid interactions. Nat Protoc 2:1849–1861
10. Unterholzner SJ, Rozhon W, Papacek M et al (2015) Brassinosteroids are master regulators of gibberellin biosynthesis in Arabidopsis. Plant Cell 27:2261–2272
11. Fried M, Crothers DM (1981) Equilibria and kinetics of *lac* repressor-operator interactions by polyacrylamide gel electrophoresis. Nucleic Acids Res 9:6505–6525
12. Senear DF, Brenowitz M (1991) Determination of binding constants for cooperative site-specific protein-DNA interactions using the gel mobility-shift assay. J Biol Chem 266:13661–13671
13. van Oeffelen L, Peeters E, Nguyen Le Minh P et al (2014) The 'Densitometric Image Analysis Software' and its application to determine stepwise equilibrium constants from electrophoretic mobility shift assays. PLoS One 9:e85146
14. Rye HS, Drees BL, Nelson HCM et al (1993) Stable fluorescent dye-DNA complexes in high sensitivity detection of protein-DNA interactions. Application to heat shock transcription factor. J Biol Chem 268:25229–25238
15. Berger R, Duncan MR, Berman B (1993) Nonradioactive gel mobility shift assay using chemiluminescent detection. Biotechniques 15(650):652
16. Rodgers JT, Patel P, Hennes JL et al (2000) Use of biotin-labeled nucleic acids for protein purification and agarose-based chemiluminescent electromobility shift assays. Anal Biochem 277:254–259
17. Ruscher K, Reuter M, Kupper D et al (2000) A fluorescence based non-radioactive electrophoretic mobility shift assay. J Biotechnol 78:163–170
18. Onizuka T, Endo S, Hirano M et al (2002) Design of a fluorescent electrophoretic mobility shift assay improved for the quantitative and multiple analysis of protein-DNA complexes. Biosci Biotechnol Biochem 66:2732–2734
19. Rozhon W, Mayerhofer J, Petutschnig E et al (2010) ASKθ, a group-III Arabidopsis GSK3, functions in the brassinosteroid signalling pathway. Plant J 62:215–223

Chapter 12

Identification of Brassinosteroid Signaling Complexes by Coimmunoprecipitation and Mass Spectrometry

Walter van Dongen, Luc van Heerde, Sjef Boeren, and Sacco C. de Vries

Abstract

A combination of coimmunoprecipitation (coIP) of tagged proteins followed by protein identification and quantitation using Liquid Chromatography Mass Spectrometry/Mass Spectrometry (LCMS/MS) has proven to be a reliable method to qualitatively characterize membrane-bound receptor complexes from plants. Success depends on a range of parameters, such as abundance and stability of the complex and functionality of the tagged receptors, efficiency of the protein complex isolation procedure, MS equipment, and analysis software in use. In this Chapter, we focus on the use of one of the green fluorescent protein-tagged receptors of the SOMATIC EMBRYOGENESIS RECEPTOR-LIKE KINASE (SERK) family, of which SERK3, also known as BRASSINOSTEROID INSENSITIVE1 (BRI1) ASSOCIATED KINASE1 (BAK1), is a coreceptor of BRI1. Like BRI1 itself, SERK3 is a leucine-rich repeat receptor kinase (LRR RK) with a single-pass transmembrane domain. The latest updated laboratory protocol is presented as well as examples of data analysis and typical results obtained. Potential drawbacks of the procedure employed for plant membrane proteins will be pointed out.

Key words Protein–protein interaction, *Arabidopsis*, Immunoprecipitation, Tandem mass spectrometry, Protein complex identification, GFP

1 Introduction

Isolation of plant receptor complexes that makes use of the availability of green fluorescent protein (GFP)-tagged proteins has several advantages over other methods. High-affinity anti-GFP antibodies immobilized on Sepharose or magnetic beads or to 96-well plates are commercially available, making pull-down procedures possible for each laboratory. Also, for large numbers of membrane receptors, *Arabidopsis thaliana* plant lines are available that express C-terminal GFP fusions [1] and http://www4.ncsu.edu/~sclouse/Clouse2010.htm. In many, but not all situations, such fusion proteins are functional [2] and semiquantitative information can be obtained concerning the precise (sub)cellular localization and abundance of the fusion proteins [3]. In the first

Eugenia Russinova and Ana I. Caño-Delgado (eds.), *Brassinosteroids: Methods and Protocols*, Methods in Molecular Biology, vol. 1564, DOI 10.1007/978-1-4939-6813-8_12, © Springer Science+Business Media LLC 2017

demonstration of the plant receptor complex by means of the immunoprecipitation method with seedlings expressing the SOMATIC EMBRYOGENESIS RECEPTOR-LIKE KINASE1 (SERK1)-GFP fusion, particular care was taken to ensure that the protein isolation method was as rapid as possible [4] Later modifications were included for both membrane-bound complexes as well as for the isolation of nuclear proteins [5], whereas, more recently, an updated procedure has been published [6]. Other prepurification strategies are also utilized, such as the use of prior microsomal membrane purification and/or removal of GFP-interacting proteins to reduce the number of false positives or specifically sticking proteins [7]. Based on our previous protocols, a number of adjustments have been introduced, primarily aimed at speeding up the procedure. Results from these faster complex isolation procedures will be presented for the isolation of the brassinosteroid (BR) signaling complex with one of the SERK family coreceptors, SERK3 or BR INSENSITIVE1-ASSOCIATED KINASE1 (BAK1)-GFP, as the tagged protein.

The protocol has been used for an extensive series of GFP-tagged proteins and has been successful and reproducible in the identification of protein complexes in approximately two-thirds of the proteins tested. Excluding trivial reasons, abundance and stability of the tagged proteins appear to be important decisive factors. Whereas membrane proteins present an additional difficulty due to the possible presence of contaminant proteins in detergent-insoluble patches of membrane, it is hard to ensure total solubilization without affecting the protein–protein interactions themselves. Experiments to use coimmunoprecipitation (coIP) followed by liquid chromatography tandem-mass spectrometry (LC-MS/MS) to identify transient ligand-induced changes in protein complex compositions have met with little success, most probably due to the relatively minor percentage of membrane protein complexes that undergo activation at any given time [8]. Finally, it should be emphasized that the identification of protein complex components requires additional lines of evidence, such as colocalization in the appropriate cellular compartments, genetic evidence, or physiological data.

2 Materials

2.1 Plant Material

Arabidopsis thaliana (L.) Heynh. seedlings from homozygous transgenic lines expressing SERK or other receptor GFP fusion proteins capable of rescuing potential mutant phenotypes. For an easily detectable receptor–GFP fusion, expressed in all cells and visible at standard setting under a confocal microscope (*see* [3] for details), between 1 and 3 g of frozen powdered 4- to 6-day-old seedling material is sufficient per single technical replicate, but three technical replicates are required per biological replicate.

2.2 Chemicals, Reagents, Equipment, and Other Materials

1. Extraction buffer (EB+): 50 mM Tris–HCl (pH 7.5), 150 mM NaCl, 1 % (v/v) nonyl phenoxypolyethoxylethanol (NP-40; currently available as Igepal CA-630; *see* **Note 1**), and tablet protease inhibitor Complete. In extraction buffer EB-, the detergent NP-40 is omitted.
2. Elution buffer: freshly prepared 50 mM $NH_4(CO_3)_2$ (pH 8.0) without protease inhibitors.
3. Dithiothreitol (DTT).
4. Iodo-acetamide (IAA).
5. L-cysteine.
6. Trypsin, sequencing grade (0.5 μg/μL in 1 mM HCl).
7. Trifluoroacetic acid (TFA).
8. Methanol.
9. Formic acid (HCOOH).
10. Acetonitrile (AcNi).
11. 14- and 50-mL Falcon disposable tubes.
12. Mortar and pestle.
13. Needle sonicator fitted with a micro tip needle probe.
14. 40-μm cell strainer.
15. Anti-GFP-coated magnetic beads (μMACS GFP isolation kit, Miltenyi).
16. Magnetic columns (μMACS; Miltenyi).
17. MultiStand (μMACS; Miltenyi).
18. μMACS Separator (Miltenyi).
19. 0.5-mL protein low-bind tubes (Eppendorf LoBind).
20. C18 Empore membrane (3M).
21. 1.6-mm tissue puncher (Hamilton Gauge 14 needle and a fitting piston).
22. LiChroprep RP-18 columns.
23. Home-made C18 μ columns (C18 membrane frit and C18 column material) (*see* Subheading 3, **step 17**).
24. A table centrifuge capable of creating a vacuum (SpeedVac).
25. LC-MS/MS LTQ-Orbitrap XL.
26. Xcalibur 2.1 software.

3 Methods

1. Thoroughly grind plant material in liquid nitrogen with mortar and pestle. Transfer the ground material to a 50-mL tube and place it in liquid nitrogen.

2. Weigh 1–3 g of plant tissue powder into a precooled tube. Make sure all tubes contain the same amount of tissue with no more than an approximately 10 % difference between tubes. Store at −80 °C.
3. Precool mortar on ice. Prepare and cool extraction buffers.
4. Add weighed material to mortar and add as little extraction buffer EB+ as possible to suspend the plant material (usually approximately 8–9 mL for 3 g material) and grind again very thoroughly with mortar and pestle.
5. Transfer the protein extract to a 14-mL tube and sonicate with a needle sonicator fitted with a micro tip probe three times at 75 % power for 15 s with at least a 15 s pause in between. Keep the samples on ice during sonication to avoid protein denaturation, but no longer than 30 min, including the time of the sonication procedure.
6. Reduce the detergent concentration to 0.2 % (v/v) NP-40 in the protein extract by adding extraction buffer EB- and transfer the protein extract to centrifuge tubes to improve the flow-through of the magnetic columns.
7. Centrifuge at 25,000 × *g* for 15 min at 4 °C. If any debris remains in the supernatant, the magnetic columns will get stuck or run very slowly; in that case, a second centrifugation run at the same settings can be introduced. This second run is routine in the standard protocol [6].
8. Transfer the supernatant through a 40-μm cell strainer into a 50-mL tube and add 50–100 μL anti-GFP magnetic beads (50 μL per gram of plant material, but not more than 100 μL). Resuspend the beads before pipetting them with a cut tip. Mix well by swirling and transfer the tube to the cold room (*see* **Note 2**).
9. Place the magnetic columns in the magnetic field of the μMACS Separator.
10. Add 200 μL extraction buffer with 0.1 % NP-40 to the column.
11. Apply protein extract with beads onto the column and let the lysate run through. Columns are "flow-stop" and do not run dry. For 25 mL of protein extract, it takes approximately 2 h. During that time, occasionally swirl the 50-mL tube.
12. Rinse the column with 4× 200 μL extraction buffer with 0.1 % NP-40 and 2× 500 μL of 50 mM NH_4HCO_3.
13. Remove the column from the magnet, immediately place a 0.5-mL Low Bind tube below the column and elute the beads into the tube with 50 μL of 50 mM NH_4HCO_3 preheated at 95 °C. Make sure that the "end" of the column is in contact with the eluate all the time. Transfer column into a new Low Bind tube and add another 50 μL of preheated hot elution buffer to the column to check whether most of the beads have been collected in the first tube. Store at −20 °C.

14. Add 1 μL of 500 mM DTT in 50 mM NH_4HCO_3 to each immunoprecipitated replicate sample at 10 mM final concentration and incubate at 60 °C for 1–2 h. Add 1 μL 750 mM IAA in 50 mM NH_4HCO_3 at 15 mM final concentration and incubate for maximally 1 h at room temperature in the dark. Add 1 μL of 200 mM L-cysteine in 50 mM NH_4HCO_3 to stop the alkylation reaction.
15. Add 1 μL trypsin sequencing grade (0.5 μg/μL in 1 mM HCl) to each immunoprecipitated sample and incubate overnight at 20 °C on a shaker. Do not incubate longer than approximately 16 h because it will increase the amount of chymotrypsinic cleavages.
16. Carefully add 1.5–3 μL 10 % (v/v) TFA to make the pH approximately pH 3. Check the pH with a pH paper.
17. The peptide solutions are now concentrated and desalted with homemade C18 μ columns (C18 membrane frit + C18 column material). With μColumns, no hydrophilic peptides will be lost. μColumns can be prepared the day before by cutting a small circular (approximately 1.6 mm^2) piece of a C18 Empore disk (=frit) and to push this into a 200-μL pipette tip with a well-fitting Hamilton plunger. Add 200 μL of methanol to the tip with frit. Prepare a 50 % (v/v) slurry of Lichroprep C18 column material and add 4 μL of the 50 % slurry into the methanol in the tip + frit. The prepared μColumns can be stored overnight and can be eluted by hand with a plastic 5-mL syringe or with a vacuum manifold. Do not let the μColumns run dry.
18. Wash the μColumns with 100 μL of MeOH, equilibrate with 100 μL of 0.1 % (v/v) HCOOH, add the acidified peptide samples to the μColumns and elute after a wash with 100 μL 0.1 % (v/v) HCOOH. If required the peptide samples can be diluted with 0.1 % (v/v) HCOOH, but make sure no acetonitrile is present.
19. After transfer of the μColumn to a 0.5-mL Low Binding tube, elute the bound peptides manually with 50 μL of 50 % (v/v) AcNi in 0.1 % (v/v) HCOOH directly into a 0.5-mL Low Binding tube. To reduce the AcNi content prior to the LC-MS/MS analysis, evaporate the samples in a SpeedVac at 30–45 °C to a small volume in approximately 2 h. The resulting volume should be below 10 μL. Adjust the sample volume with 0.1 % (v/v) HCOOH in water to 50 μL, sonicate several times for 10 s, centrifuge at room temperature at 3000 × *g*, and store at −20 °C.
20. Separate peptides by reversed-phase nano liquid chromatography connected to the MS via a nanoelectrospray interface and measure MS spectra online by accurate MS as well as by collision-induced dissociation-fragmented MS/MS spectra. Our data are produced by an LTQ-Orbitrap XL with Xcalibur 2.1 software (for further details for this particular machine, *see* Tables 1 and 2 of ref. [6]).

21. Analyze all LC-MS/MS runs obtained with all MS/MS spectra by means of the database search algorithm Andromeda from MaxQuant in the "Label Free Quantification" mode (for further details on these settings, *see* Table 3 of ref. [6]). Normalized quantitative information (Label-Free Quantification [LFQ] intensity) is obtained for all protein groups identified when two or more peptides could be properly quantified. The Perseus filtering and statistics software can be used to filter out the 1 % reversed hits and to delete the proteins that are identified in less than half of the replicates (for further details on the procedure, *see* Table 4 of ref. [6]).
22. Create groups of replicates and perform Student's *t*-tests between groups in Perseus, providing not only the *P* value (column: - Log *t*-test *P* value), but also immediately yielding the ratio of the group average Log LFQ values (column: *t*-test Difference) when applied to two groups. Significance at a variable false discovery rate (FDR) is judged by both *P* value and ratio, of which the weight is set by a variable S0 value. Both FDR and S0 should be optimized to yield no or only a few proteins significantly different on the control side of the volcano plot obtained (Fig. 1).
23. In a successful experiment, the bait (tagged) protein should be in the top 10 when the list is sorted by ratio (Table 1). A high ratio and low *P* value should make the difference clearly different between sample and control.

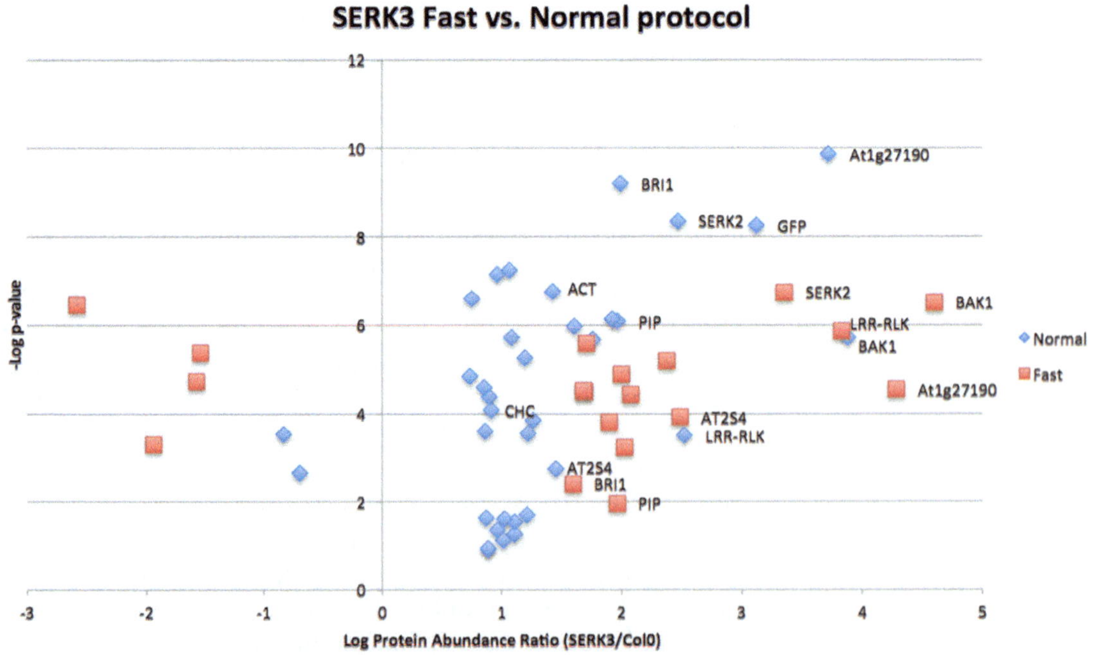

Fig. 1 Typical output graphs of an immunoprecipitation experiment with SERK3/BAK1-GFP seedlings and comparison of the fast (*red squares*) with the standard (*blue diamonds*) protocol [6]. Contaminants have been excluded. Relevant parts of the corresponding data sheet are shown in Tables 1 and 2

Table 1

Summary of the top hits found in both the standard [6] and fast protocols (described here)

Gene	Position after sorting by protein abundance ratio		Log protein abundance ratio		−Log *P* value					Intensity
Gene	Standard	Fast	Standard	Fast	Standard	Fast	Standard	Fast	*BAK1/SERK3*	1
1	3.9	4.6	5.727346	6.512613	1.02E + 10	4.78E + 09			*At1g27190*	2
2	3.7	4.3	9.870817	4.541265	6.50E + 09	3.03E + 09			*LRR-RLK*	4
3	2.5	3.8	3.502851	5.862078	5.14E + 08	7.89E + 08			*SERK2*	5
4	2.5	3.4	8.35192	6.745251	3.61E + 08	2.61E + 08			*AT254*	11
5	14	2.5	2.739587	3.911273	1.19E + 09	1.88E + 08		*CRU3;CRC*	26	8
0.9	2.0	4.373926	3.224458	5.41E + 07	4.20E + 07			*PIP1B;PIP1-2*	7	10
2.0	2.0	6.086577	1.959826	2.58E + 08	2.03E + 08			*BRI1*	6	17
2.0	1.6	9.202629	2.407155	5.30E + 10	2.60E + 10					

Data are derived from the output files of ref. [6]. The log protein abundance ratios for the same protein differ between both protocols, partly due to the fact that the experiments were carried out with their own internal controls on different days. Whereas the top four or five proteins differ very little in position, clearly less prevalent proteins can be quite different in ranking between the normal and fast protocols. Whether this reflects a meaningful biological variable is presently unknown

Table 2
Read-outs of SERK3-GFP pull-downs using the standard [6] and the fast protocols used to draw Fig. 1

Fast			Standard (ref. [6])		
Gene	Ratio	Significant	Gene	*t*-test Difference_ SERK3 vs wild type	Significant
BAK1	4.600310326	6.512612909	*BAK1*	3.882249355	5.727346316
At1g27190	4.280583858	4.541264623	*At1g27190*	3.717175007	9.870816622
LRR-RLK	3.828957876	5.862078021	*GFP*	3.125950654	8.252725038
SERK2	3.358774503	6.745251295	*LRR-RLK*	2.527445316	3.502851291
AT2S4	2.492005507	3.911272809	*SERK2*	2.470540841	8.351919808
AT2S3	2.373407523	5.181272835	*BRI1*	1.990800222	9.202628669
At4g36700	2.077480634	4.434374653	*PIP*	1.968911171	6.086576639
CRU3	2.022760073	3.224457615	*AHA1*	1.92390728	6.132879176
METK4	1.997889042	4.898635312		1.764056206	5.674252248
PIP	1.965910594	1.959825537	*AAC1*	1.609477361	5.966836978
psbE	1.89700826	3.805590606	*AT2S4*	1.447920163	2.739587152
RPS24B	1.705882867	5.578831975	*ACT*	1.427423477	6.74972137
CPN60	1.693805536	4.501740571		1.265810013	3.843305308
PEPR1	1.684269587	4.520620933		1.217849572	3.549997527
BRI1	1.595747471	2.407154782		1.212004503	1.708999141
GASA1	−1.534285227	5.366774299	*TUBB*	1.189728578	5.25840623
At5g07090	−1.567441463	4.730851042	*GAPC*	1.107603232	1.260547727
NUCL1	−1.933625539	3.286824673		1.107351939	1.535053411
At2g29210	−2.579381307	6.469190957		1.08503739	5.711913229
				1.061819553	7.234157636
				1.018186728	1.623476139
				1.01233228	1.125481728
				0.96467336	7.143577281
				0.959725539	1.369592406
			CHC	0.91131258	4.070918848
				0.889634609	4.373925911
				0.889151573	0.931156273
				0.881804307	0.934821863

Table 2 (continued)

Fast			Standard (ref. [6])		
Gene	Ratio	Significant	Gene	*t*-test Difference_ SERK3 vs wild type	Significant
				0.869459152	1.642004307
				0.855745474	3.596612678
				0.850540797	4.586182597
				0.746487459	6.594291388
				0.729589303	4.835408861
				−0.688692252	2.661315912
				−0.827438672	3.536885373

4 Notes

1. Care should be taken in selecting the right detergent; NP-40 differs from Nonidet P40. Other detergents tested are ß-octylglucoside, resulting in increased background and Triton X-100, yielding results similar to NP-40.
2. In the standard protocol [6], the diluted protein extract (Subheading 3, **step 6**) is incubated for 3 h in the cold room, but in the modified fast protocol described here this is reduced to 30 min.

Acknowledgements

We thank Jos Wendrich (Biochemistry, Wageningen University, Netherlands) and Bert de Rybel (VIB-Ghent University, Belgium) for providing the updated immunoprecipitation protocols currently in use.

References

1. Gou X, He K, Yang H et al (2010) Genome-wide cloning and sequence analysis of leucine-rich repeat receptor-like protein kinase genes in *Arabidopsis thaliana*. BMC Genomics 11:19
2. Ntoukakis V, Schwessinger B, Segonzac C et al (2011) Cautionary notes on the use of C-terminal BAK1 fusion proteins for functional studies. Plant Cell 23:3871–3878
3. van Esse GW, Westphal AH, Surendran RP et al (2011) Quantification of the Brassinosteroid Insensitive1 receptor in planta. Plant Physiol 156:1691–1700
4. Karlova R, Boeren S, Russinova E et al (2006) The *Arabidopsis* SOMATIC EMBRYO GENESIS RECEPTOR-LIKE KINASE1 protein complex includes BRASSINOSTEROID-INSENSITIVE1. Plant Cell 18:626–638

5. Smaczniak C, Li N, Boeren S et al (2012) Proteomics-based identification of low-abundance signaling and regulatory protein complexes in native plant tissues. Nat Protoc 7:2144–2158
6. Wendrich JR, Boeren S, Möller BK et al (2017) In vivo identification of plant protein complexes using IP-MS/MS. In: Kleine-Vehn J, Sauer M (eds) Plant hormones, Methods in molecular biology, 3rd edn. Humana Press, Totowa, NJ 1497:147–158
7. Fàbregas N, Li N, Boeren S et al (2013) The BRASSINOSTEROID INSENSITIVE1-LIKE3 signalosome complex regulates *Arabidopsis* root development. Plant Cell 25:3377–3388
8. Bücherl C, van Esse GW, Kruis A et al (2013) Visualization of BRI1 and BAK1(SERK3) membrane receptor heterooligomers during brassinosteroid signaling. Plant Physiol 162: 1911–1925

Chapter 13

Simplified Enrichment of Plasma Membrane Proteins from *Arabidopsis thaliana* Seedlings Using Differential Centrifugation and Brij-58 Treatment

Carina A. Collins, Michelle E. Leslie, Scott C. Peck, and Antje Heese

Abstract

The plasma membrane (PM) forms a barrier between a plant cell and its environment. Proteins at this subcellular location play diverse and complex roles, including perception of extracellular signals to coordinate cellular changes. Analyses of PM proteins, however, are often limited by the relatively low abundance of these proteins in the total cellular protein pool. Techniques traditionally used for enrichment of PM proteins are time consuming, tedious, and require extensive optimization. Here, we provide a simple and reproducible enrichment procedure for PM proteins from *Arabidopsis thaliana* seedlings starting from total microsomal membranes isolated by differential centrifugation. To enrich for PM proteins, total microsomes are treated with the nonionic detergent Brij-58 to decrease the abundance of contaminating organellar proteins. This protocol combined with the genetic resources available in *Arabidopsis* provides a powerful tool that will enhance our understanding of proteins at the PM.

Key words Plasma membrane, Receptor-like kinases (RLKs), Protein enrichment, Ultracentrifugation, Differential centrifugation, Cell fractionation, Brij-58 detergent, *Arabidopsis*, Seedlings

1 Introduction

Plants live in a constantly changing environment and must be able to both rapidly perceive and respond to diverse extracellular stimuli. To detect changes in the environment and initiate intracellular responses, plants have an extensive suite of proteins localized at the plasma membrane (PM). The largest class of plant PM-associated signaling proteins is the receptor-like kinases (RLKs). These RLKs are Type-I membrane proteins containing various extracellular domains involved in signal perception, a single transmembrane domain, and an intracellular kinase domain for signal initiation [1]. One well-studied RLK is BRASSINOSTEROID-INSENSITIVE1 (BRI1), which along with its coreceptor BRI1-ASSOCIATED KINASE1 (BAK1) promotes plant growth through perception of brassinosteroid hormones [2–7]. In addition to roles in signaling

Eugenia Russinova and Ana I. Caño-Delgado (eds.), *Brassinosteroids: Methods and Protocols*, Methods in Molecular Biology, vol. 1564, DOI 10.1007/978-1-4939-6813-8_13, © Springer Science+Business Media LLC 2017

for growth, PM-localized RLKs are required for pathogen perception [8]. For example, the RLK FLAGELLIN SENSING2 (FLS2) recognizes bacterial flagellin (or its active peptide-derivative flg22) to initiate immune responses [9, 10]. Interestingly, similar to BRI1, FLS2 rapidly associates with BAK1 after ligand perception, and this complex formation is required for signal initiation [11, 12]. Beyond signaling, other PM proteins have essential functions in cell wall remodeling [13–15], such as the cellulose synthase complexes that deposit newly synthesized polysaccharides to build and maintain the integrity of the cell wall [13–16]. Furthermore, additional proteins at the PM are necessary for nutrient transport across the membrane [17, 18]. Although roles for many PM proteins have been identified based on genetics, the biochemical characterization of these proteins (and their complexes) remains an important area of investigation.

Study of PM-associated proteins in the laboratory is a challenge, because these proteins are hydrophobic and often present in low abundance relative to the total cellular protein content [19–23]. To enrich for PM proteins, the PM must be separated not only from the cytosol, but also from other organelles, such as from the endoplasmic reticulum (ER), mitochondria, tonoplast, and chloroplasts. Traditionally, isolation of the PM is accomplished by aqueous two-phase partitioning. This technique separates membranes (and associated proteins) based on membrane surface properties as opposed to size or density [24]. Subtle differences in experimental parameters of two-phase partitioning can dramatically alter the outcome of this technique. Therefore, tedious optimization of the protocol is often required for each species and each tissue from the same species [24, 25]. Development of a simplified and robust method that can be used across plant species and tissue types will facilitate additional studies leading to a better understanding of PM protein functions.

Here a simplified method is presented to enrich for PM proteins by way of removing contaminating organellar proteins. During the initial step of protein extraction (homogenization of plant tissue), the PM (as well as the ER, tonoplast, and other organellar membranes) is broken into small membrane fragments that spontaneously collapse into vesicles. During this process, organellar membranes and cytosolic contents may become trapped within these PM-derived vesicles. Following tissue homogenization, extracts are first cleared by low-speed centrifugation (i.e. 8000 × *g*) to remove intact cells, cell wall debris, and intact organelles, including mitochondria and nuclei. Ultracentrifugation of this postmitochondrial fraction is used to isolate microsomes that contain primarily PM, ER, Golgi apparatus, tonoplast, and endomembranes (Fig. 1).

Treatment of these microsomes with the nonionic detergent Brij-58 inverts PM-derived vesicles [26], theoretically contributing

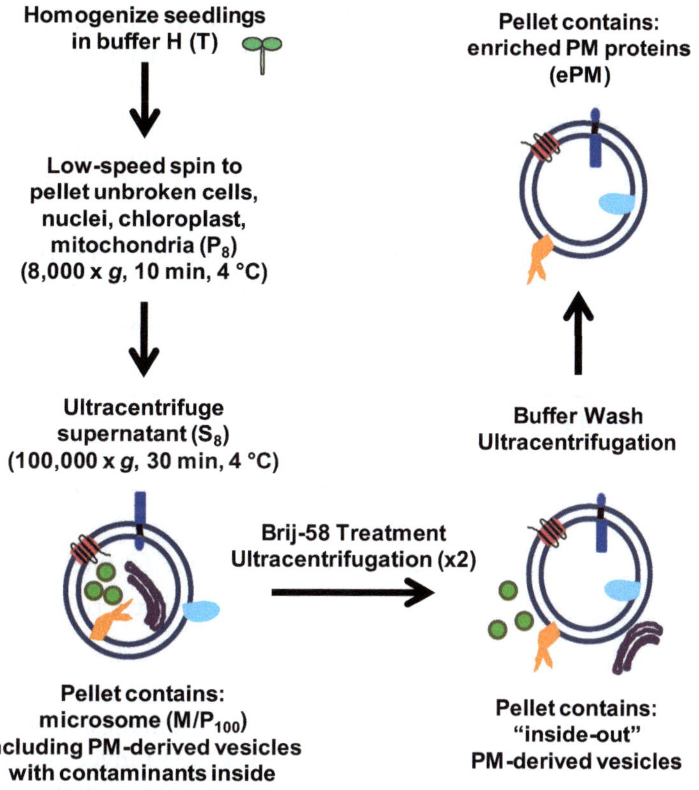

Fig. 1 Schematic representation of the simplified PM enrichment procedure. After homogenization of the *Arabidopsis* seedlings in Buffer H, the total cell homogenate (T) is subjected to a low-speed spin (8000 × *g*) to separate the supernatant (S_8; containing cytosolic, soluble, and microsomal proteins) from the pellet (P_8; containing cell debris and large organelles). The microsomal membranes (M/P_{100}) are then separated from cytosolic and soluble proteins (S100) by ultracentrifugation (100,000 × *g*). After treatment with the detergent Brij-58, PM vesicles are inverted to release contaminating contents. Pellets are washed twice with Buffer H and subjected to a final ultracentrifugation step to yield an enriched PM (ePM) fraction, depleted of other organellar membrane proteins

to the release of contaminating contents of vesiculated PMs. Although the method has been clearly demonstrated to enrich for PM proteins, it is difficult to conclusively prove the mechanism by which Brij-58 performs its function. In addition to the inversion of PM-derived vesicles, Brij-58 may selectively solubilize ER-derived membranes [27]. Thus, we cannot exclude the possibility that, in addition to releasing contaminants, the use of Brij-58 may also allow for the selective solubilization of other (non-PM) membranes. Irrespectively, Brij-58 treatment results in a fraction that is enriched in PM proteins and decreased in the abundance of other membrane proteins. Originally established for *Arabidopsis thaliana* suspension cell cultures [23], the PM enrichment method was successfully applied to maize (*Zea mays*) root tissue [28]. Here, we adapted this method to *Arabidopsis* seedlings, indicating that this is a robust procedure that can be utilized for PM protein enrichment from different tissue types with minimal optimization. The ability to combine the PM enrichment protocol in seedlings with the large number of genetic resources available in *Arabidopsis* will allow researchers to answer diverse biological questions, thus providing an important step toward furthering our functional understanding of PM proteins and their regulation.

2 Materials

1. Six- or 7-day old *Arabidopsis thaliana* (L.) Heynh. seedlings (*see* **Note 1**).
2. Transparent, sterile 12-well tissue-culture plate.
3. Sterile, deionized water (dH_2O) at room temperature.
4. Forceps.
5. Buffer H: 250 mM sucrose, 50 mM N-(2-hydroxyethyl) piperazine-N′-(2-ethanesulfonic acid) (HEPES)-KOH (pH 7.5), 5 % (v/v) glycerol, 50 mM sodium pyrophosphate decahydrate, 1 mM sodium molybdate dihydrate, 25 mM sodium fluoride, 10 mM ethylenediamine tetraacetic acid, tetrasodium salt dehydrate (EDTA), 0.5 % (w/v) polyvinylpyrrolidone (PVP-10) with the following inhibitors added to a final concentration of 3 mM dithiothreitol (DTT), 1 mM phenylmethanesulfonyl fluoride (PMSF), 10 μM leupeptin, and 10 nM calyculin A (*see* **Note 2**). Keep on ice.
6. 5× polyethylene glycol hexadecyl ether (Brij-58) solution (0.1 %): Dissolve 0.01 g Brij-58 in 10 mL Buffer H. Once Brij-58 is evenly resuspended, dilute 5× Brij-58 stock solution to 1× with Buffer H (from 0.1 % to a final concentration of 0.02 %; *see* **Note 3**). A final volume of 10 mL 1× Brij-58 is sufficient for a single experiment. Keep on ice.
7. Kimwipes or delicate task wipers.
8. Laboratory grade quartz sand (*see* **Note 4**).
9. 50- and 15-mL conical tubes.
10. Timer.
11. Mortar and pestle (*see* **Note 5**).
12. Miracloth, cut into 10 × 10 cm squares.
13. Microcentrifuge tubes (1.7 and 2.0 mL).
14. Tabletop microcentrifuge with temperature control (*see* **Note 6**).

15. Preparative tabletop ultracentrifuge with a TLA 100.2 fixed-angle rotor fitted with thickwall polycarbonate tubes (7 × 21 mm) or similar, such as Optima TXL-120 (Beckman Coulter).
16. Long polypropylene pestles with rounded nonhomogenization ends, such as 21 cm length × 4.8 mm shaft conical tissue grinders.
17. Bradford Reagent.
18. Spectrophotometer with 1-mL disposable cuvettes.
19. Sodium dodecyl sulfate-polyacrylamide gel electrophoresis (SDS-PAGE) equipment and materials.
20. Transfer equipment and materials for immunoblot analysis.
21. Antibodies against subcellular marker proteins (*see* **Note** 7). Our laboratory has had success with the following antibodies from Agrisera (Vännäs, Sweden) made against the following proteins: PM-localized H^+-ATPase (AHA), integral ER membrane sterol methyl transferase (SMT1), ε subunit of tonoplast H^+-ATPase (V-ATPase), and outer envelope membrane protein (OEP75) from chloroplast translocon complex (Toc75).
22. Ponceau S stain solution (*see* **Note** 7).

3 Methods

3.1 Total Protein Isolation from Arabidopsis Seedlings

A mortar and pestle are used to homogenize whole *Arabidopsis* seedlings for total protein isolation. Homogenate is filtered to remove the majority of cell debris.

1. Approximately 16–20 h before starting to grind the tissue, transfer 180–200 6- or 7-day-old *Arabidopsis* seedlings (approximately 500–700 mg fresh weight; *see* **Note 8**) per sample to a 12-well tissue culture plate. Evenly distribute the seedlings for each sample across approximately four wells, each containing 3 mL sterile dH_2O (Fig. 2a, b). Place the plate in continuous light at 22 °C overnight (*see* **Note 9**).
2. On the day of the experiment, chill mortar and pestle in a bucket of ice. Add 1.5–2.5 mL of freshly prepared Buffer H to the chilled mortar along with 0.1 g of sand (*see* **Note 4**). From this point on, all steps must take place on ice or at 4 °C.
3. Using forceps, gently, but swiftly, remove all seedlings within a genotype or treatment from the 12-well plate (*see* **Notes 9** and **10**). Gently blot excess water from the seedlings on a Kimwipe before transfer to the chilled mortar.
4. Grind seedlings with the chilled mortar and pestle for at least 3 min or until completely homogenized (*see* **Note 11**).

Fig. 2 Grinding of *Arabidopsis* seedlings and filtering of protein homogenate. (**a**) Six- to 7-day-old wild-type (Col-0) *Arabidopsis* seedlings typically used for an experiment. (**b**) Seedlings from (**a**) placed into wells of a 12-well plate containing 3 mL of sterile dH_2O. (**c**) Two layers of Miracloth placed into a 15-mL conical tube with the aid of the back of a permanent marker. (**d**) After grinding of the seedlings for the appropriate amount of time, transfer of the homogenate to a conical tube lined with Miracloth. (**e**) To pass homogenate through the Miracloth, one finger used to push the Miracloth against the tube side, while twisting the Miracloth with the other hand, thus helping to force the liquid through the Miracloth. (**f**) After squeezing, through the Miracloth, the homogenate is collected in the bottom of the tube

5. After lining the opening of a 15-mL conical tube with two layers of Miracloth squares, put cell homogenate into and squeeze/twist it through the Miracloth to filter out debris (Fig. 2c–f; *see* **Note 12**).
6. Rinse mortar and pestle with an additional 0.5–1.0 mL of Buffer H.
7. Collect, rinse, and add to the content of the 15-mL conical tube used in **step 5** by filtering and squeezing the additional solution through the same Miracloth as before (*see* **Notes 12** and **13**). This fraction represents the cleared homogenate or total (T) protein.
8. Save approximately 50 μL of T for immunoblot analysis. Keep in labeled microcentrifuge tube on ice.
9. If applicable, repeat **steps 2–8** for homogenization of additional samples.

3.2 Microsomal Protein Isolation

First, large organelles, including nuclei, mitochondria, and most chloroplasts, are removed from the total protein (T) by a low-speed centrifugation step. The supernatant (S_8) containing cytosolic, soluble, and microsomal proteins is then subjected to ultracentrifugation to separate microsomal proteins (M = P_{100}) from cytosolic and soluble proteins (S_{100}).

1. Transfer filtered homogenate (T) from 15-mL conical tubes to 2.0-mL microcentrifuge tubes for brief centrifugation (8000 × *g* for 10 min at 4 °C; *see* **Note 6**). This low-speed centrifugation step will pellet any remaining nonhomogenized cellular debris and remove nuclei, mitochondria, and most of the chloroplasts (referred to as P_8), thus separating P_8 and S_8.
2. Transfer supernatant (S_8), which contains cytosolic proteins, microsomal proteins, and soluble organellar proteins (such as ribulose-1,5-bisphosphate carboxylase/oxygenase [RuBisCo]) released during cell lysis, to ultracentrifuge tubes. Place tubes in appropriate rotor and begin first ultracentrifugation step (100,000 × *g* for 30 min at 4 °C).
3. Gently remove tubes from rotor to avoid disturbing pellet. At this stage, the supernatant (S_{100}) contains soluble and cytosolic proteins, while the pellet (P_{100} = M) contains membrane proteins from the PM, ER, tonoplast, Golgi as well as some contaminating membrane and soluble organellar proteins (such as chloroplasts and RuBisCo).
4. Carefully transfer S_{100} from each ultracentrifuge tube to a new microcentrifuge tube, saving approximately 50 μL of S_{100} for immunoblot analysis. Keep in labeled microcentrifuge tubes on ice.
5. Remove and discard all remaining supernatant from ultracentrifuge tubes, taking care not to disturb pellets.
6. For each sample, resuspend M (P_{100}) in 20 μL Buffer H using the rounded end of the long pestle (Fig. 3a–d; *see* **Note 14**). Slowly add more Buffer H in 20–40 μL increments to an approximate total volume of 100 μL (*see* **Note 15**). Alternatively, sonication can be used to resuspend the pellet (*see* **Note 16**).
7. Save 20 μL of each resuspended M (P_{100}) for immunoblot analysis. Keep in labeled microcentrifuge tubes on ice.
8. Determine the protein concentration of all fractions collected at this point (S_8, S_{100}, and M) using an appropriate assay. Our laboratory typically uses Bradford protein determination assays.

3.3 Enrichment for PM Proteins

The microsomal fraction (M) isolated in the Subheading 3.2 is treated with Brij-58 detergent to invert PM-derived vesicles followed by ultracentrifugation. This step releases any contaminating organellar proteins trapped within these vesicles into the soluble fraction, thus allowing for PM protein enrichment (Fig. 1).

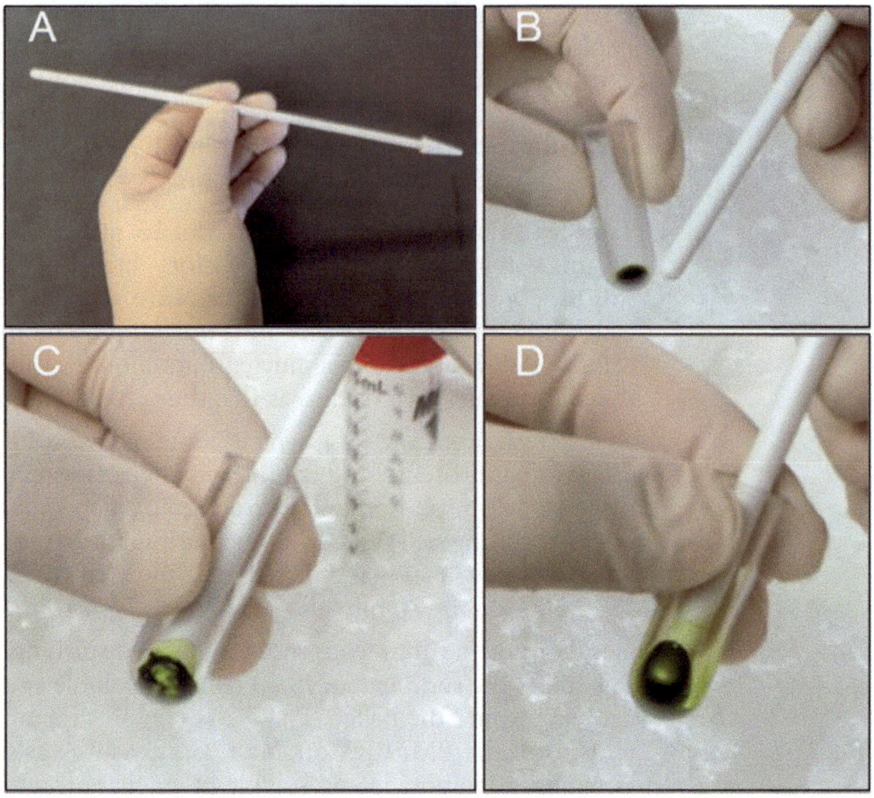

Fig. 3 Resuspension of the microsomal pellet with the rounded end of a long pestle. (**a**) Example of the type of pestle used for resuspension (*see* **Note 8**). (**b**) Close-up of the rounded end of the pestle and of the microsomal pellet after the first ultracentrifugation spin. (**c**) Initial resuspension of the pellet in 20 μL of Buffer H. The pestle is held like a pencil and the pellet is gently resuspended by circular motions against the side of the ultracentrifuge tube. (**d**) Example of the pellet after complete resuspension with the pestle

1. The target calculation is to reach a final volume of 2 μL of 1× Brij-58 (i.e., 0.02 % Brij-58 in Buffer H) per μg of total protein for each M fraction (*see* **Note 17**). Based on protein concentrations (determined in Subheading 3.2, **step 8**), first determine the necessary final volume. To the resuspended M, add 1/5th of the final volume of 5× Brij-58 solution and then immediately bring up to the final volume with Buffer H (without Brij-58). Mix by gently pipetting (but avoid creating bubbles) and incubate M fractions with Brij-58/Buffer H on ice for 30 min in ultracentrifugation tubes.
2. After the 30-min incubation, perform ultracentrifugation (100,000 × *g* for 30 min at 4 °C).
3. After carefully removing tubes from rotor, discard the supernatant. Be careful to avoid disturbing pellets.
4. Resuspend pellets as done previously (Subheading 3.2, **step 6**), but this time use 1× Brij-58 for the resuspension. First, resuspend

pellets with 20 μL of 1× Brij-58 using a rounded pestle (*see* **Notes 14** and **15**). Add 1× Brij-58 in 20–40 μL increments up to the same final volume as done for **step 1**. Incubate samples on ice for 30 min (*see* **Note 18**).

5. Repeat ultracentrifugation (100,000 × *g* for 30 min at 4 °C).
6. After centrifugation, carefully remove the supernatant, taking care not to disturb the pellet.
7. As an additional wash step, resuspend the pellet in 1 mL Buffer H (without Brij-58) as earlier (*see* **Note 19**).
8. Repeat ultracentrifugation (100,000 × *g* for 30 min at 4 °C).
9. After centrifugation, carefully remove the supernatant, taking care not to disturb the pellet.
10. Resuspend the final pellet using a rounded pestle as done for earlier steps. Begin by resuspending the pellet in 20 μL Buffer H (without Brij-58), then gradually add more Buffer H (without Brij-58) in 20–40 μL increments to a final volume of approximately 80–100 μL (*see* **Note 20**). This final, resuspended pellet is the "enriched plasma membrane" (ePM) fraction.
11. Determine the final protein concentration of the ePM fraction as done in Subheading 3.2, **step 8**.
12. Adjust the concentration of all fractions (S_8, S_{100}, M, and ePM) by adding Buffer H (without Brij-58) to a final protein concentration of 2 μg/μL. Add an appropriate amount of an SDS-containing loading buffer for immunoblot analysis (*see* **Note 21**).
13. Perform immunoblot analysis to determine whether the protein of interest is present in the ePM fraction. For optimal protein detection, load at least 40 μg of each fraction of interest onto gels. In addition, to allow for proper comparison and correct interpretation between samples, make sure to load equal concentrations (such as 40 μg) of each fraction (e.g., S_{100}, M (P_{100}), and ePM). It is crucial to confirm the quality of the PM protein enrichment by immunoblot analyses for each experiment by means of appropriate antibodies against marker proteins (Fig. 4; *see* **Notes 7** and **22**). ePM fractions can be utilized for subsequent analysis with mass spectrometry (*see* **Note 23**).

4 Notes

1. For consistent results, our laboratory grows sterilized *Arabidopsis* seeds on 0.6 % (w/v) agar plates with half-strength Murashige and Skoog (MS) salts and 1 % (w/v) sucrose for 6–7 days under continuous light at 22 °C [12, 29, 30].
2. Filter-sterilized Buffer H without inhibitors (DTT, PMSF, leupeptin, and calyculin) can be made ahead of time and stored at 4 °C.

On the day of the experiment, transfer **30** mL of Buffer H to a new 50-mL conical tube and add the inhibitors on the day of use. The volume of Buffer H will be sufficient for the homogenization of several samples of seedling tissue as well as for making 5× and 1× Brij-58 solutions.

3. The amount of Brij-58 in the final 1× stock solution (0.02 % [v/v]) is very small, so it is best to make up a 5× stock solution in Buffer H (0.1 % [v/v]) and dilute to 1× in Buffer H as needed. When making solutions, mix gently to avoid foaming of detergent. Brij-58 may take a few minutes to go into solution.
4. We found that the addition of sand can help to improve the homogenization process and to release more of the total protein content. The addition of sand, however, is optional.
5. For a typical experiment, we use a porcelain mortar with a capacity of 50 mL and a porcelain pestle with a flat bottom.
6. We use a tabletop microcentrifuge placed within a 4 °C deli case fridge.
7. It is important that for each experiment, antibodies against marker proteins of subcellular compartments are used to assess the quality of the PM protein enrichment. Consistent with being a known PM protein [23], AHA is enriched in the ePM fraction compared to the M (P_{100}) but is depleted from the cytosolic (S_{100}) fractions (Fig. 4). SMT1, Toc75, and V-ATPase are all present in the M fraction and depleted in the ePM fraction (Fig. 4). By staining the membrane with the reversible dye Ponceau S, we find that RuBisCo, a highly abundant and soluble protein present in the chloroplast stroma, is also depleted from the ePM fraction (Fig. 4).
8. This PM enrichment works well with the amount of tissue described, but, if necessary, it can be easily scaled up. Simply increase the amount of starting tissue as needed and increase the volume(s) of Buffer H proportionally for the initial grinding and rinsing steps. However, be careful not to increase the amount of buffer too much, because it may dilute the initial protein concentration.
9. It is easiest to transfer seedlings out of the 12-well plates and homogenize them when they are first placed into dH_2O. Additionally, this helps to reduce any wounding induced by the seedling removal from the MS plate. Furthermore, PM enrichment can be combined with treating seedlings with a biotic or abiotic stimulus. Placing the seedlings in dH_2O allows easy exchange of water for the solution containing the hormone, peptide, or compound of interest (*see* **Note 10**).
10. PM enrichment can be performed using naïve tissue from different *Arabidopsis* genotypes. In addition, this procedure can

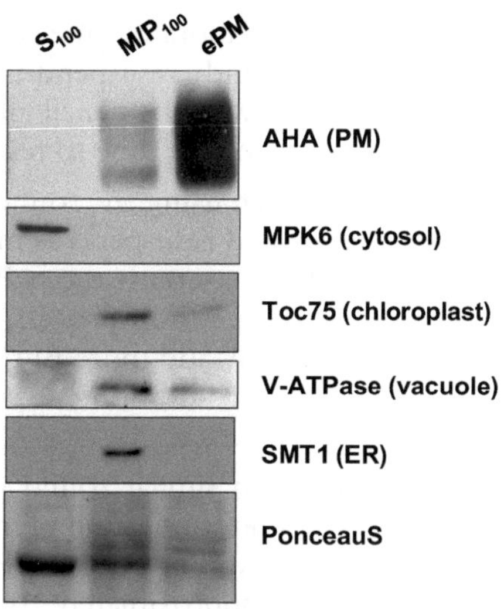

Fig. 4 Representative immunoblot for PM protein enrichment with antibodies against subcellular marker proteins. The PM marker AHA is enriched in the ePM fraction compared to the microsomal fraction (M/P_{100}). The cytosolic marker MPK6 is only present in the soluble fraction (S_{100}), consistent with a lack of cytosolic contamination in the membrane fractions. The membrane markers for chloroplast, vacuolar, and ER proteins Toc75, V-ATPase, and SMT1, respectively, are present in the microsomal fraction but reduced from the ePM fraction. These results provide proof of concept that this method allows enrichment of PM proteins, while decreasing contaminating organellar proteins, from *Arabidopsis* seedlings

be used to assess changes in the PM composition after treatment with specific hormones (such as brassinosteroids) or other abiotic or biotic compounds.

11. For reproducibility, keep grinding times and force consistent between samples within an experiment as well as between experiments (for instance, set timer to 3 min for each sample, grinding tissue for the entire time).
12. Miracloth filters tissue debris before the initial centrifugation step. It works best to "stuff" the layers of the Miracloth into the tube using a finger or a pen (Fig. 2c). Once the homogenate is added into the Miracloth, it will not drip through the material. Instead, you must squeeze/twist it through.
13. Total homogenate should be relatively close to the initial amount of Buffer H. For example, when grinding with 1.5 mL of Buffer H and rinsing with 0.5 mL of Buffer H, total homogenate should be approximately 2.0 mL (as measured in 15-mL tubes).

14. Our laboratory uses the rounded end of a long pestle (Fig. 3a) to gently resuspend the microsomal pellet. Hold the pestle as you would a pencil and use gentle circular motions along the side of the tube to resuspend the pellet.
15. When using the rounded end of a pestle, add 20 μL of Buffer H and resuspend as much of the pellet as possible. Due to its lipid content, the pellet will appear "sticky" and during the resuspension process, it will come apart in pieces (Fig. 3c). Slowly add more Buffer H, resuspending the pellet with each buffer addition. Overall, this will facilitate resuspension of the pellet.
16. When similar round pestles, such as shown here, cannot be found, sonication of microsomal pellets is an appropriate alternative. For resuspension of microsomal pellets by sonication, add approximately 100 μL of Buffer H to the pellet-containing ultracentrifuge tube. Use a small piece of parafilm to cover the tube and hold the tube in a sonicating water bath for 3–5 s intervals. Continue sonication pulses until the pellet is completely resuspended. Be careful to keep tubes on ice in between bursts of sonication.
17. For reproducibility between different experiments and between samples within the same experiment, it is crucial to keep the protein–lipid–detergent ratio constant. Here, protein concentration is used as a proxy for lipid content of the samples. We have previously determined empirically that 2 μL 0.02 % Brij-58 per μg of M proteins works best for PM enrichment [23], because too much or too little Brij-58 detergent is detrimental and results in inefficient PM enrichment. The total protein content for each microsomal sample may differ, therefore the final volume of 1× Brij-58 for each sample will vary.
18. When using whole *Arabidopsis* seedlings, this second Brij-58 incubation further decreases contaminating organellar proteins, thus increasing enrichment of PM proteins.
19. This wash step further removes any remaining cytosolic protein contamination. Importantly, it also serves to eliminate the detergent Brij-58. Detergent removal is crucial for subsequent sample analysis by mass spectrometry.
20. The final volume of the ePM fraction may vary slightly. Be careful because resuspending the pellet in too much buffer will dilute the protein concentration. It is often better to be cautious and resuspend the final pellet in less buffer to ensure a sufficient protein concentration for subsequent analyses (such as immunoblot analysis).
21. After the addition of the sample buffer, samples can be stored at −20 °C for up to two weeks. Membrane protein preparations tend to degrade when stored for longer time periods.

When protein samples will be analyzed on different days by different analyses (such as mass spectrometry and immunoblot analyses), we prepare several aliquots of each fraction to avoid repeated freeze/thaw/heat cycles, because this may contribute to protein degradation and/or aggregation.

22. Briefly for immunoblot analysis [12, 29, 30], our laboratory heats protein samples at 65 °C for 5–7 min before loading samples on 8 % (w/v) polyacrylamide gels in running buffer (0.025 M tris(hydroxymethyl)aminomethane [Tris] base, 0.192 M glycine, 0.1 % [w/v] SDS). Avoid boiling samples, because PM proteins tend to aggregate at very high temperatures and may not enter the running gel of the SDS-PAGE. Gels are run until the 35-kDa marker is near the bottom of the gel. Proteins are transferred to a 0.45-μm nitrocellulose membrane on ice (or at 4 °C). The following transfer buffer works well for large membrane proteins: 0.025 M Tris base, 0.192 M glycine, 0.025 % (w/v) SDS, 20 % (v/v) methanol, pH 8.3. Membranes are blocked with 5 % (w/v) nonfat dry milk in a 1× phosphate-buffered saline solution with 0.1 % (v/v) polyoxyethylenesorbitan (Tween 20) (PBST) for 45 min and then incubated on a rocker with primary antibodies at the appropriate concentrations overnight at 4 °C. Membranes are washed in 1× PBST (each wash approximately 7–10 min, five washes in total) and incubated with secondary antibodies for 2.5 h. Wash membranes as before and develop.
23. Our laboratory has also determined that these enriched PM protein fractions are suitable for mass spectrometry analysis [23, 28].

Acknowledgements

This work was supported by grants received from the National Science Foundation to A.H. (NSF-IOS-1147032) and S.C.P. (NSF-IOS-1025837; NSF-IOS-1051286).

References

1. Shiu SH, Bleecker AB (2001) Receptor-like kinases from *Arabidopsis* form a monophyletic gene family related to animal receptor kinases. Proc Natl Acad Sci USA 98:10763–10768
2. Kinoshita T, Caño-Delgado A, Seto H et al (2005) Binding of brassinosteroids to the extracellular domain of plant receptor kinase BRI1. Nature 433:167–171
3. Hothorn M, Belkhadir Y, Dreux M et al (2011) Structural basis of steroid hormone perception by the receptor kinase BRI1. Nature 474:467–471
4. She J, Han Z, Kim T-W et al (2011) Structural insight into brassinosteroid perception by BRI1. Nature 474:472–476
5. Irani NG, Di Rubbo S, Mylle E et al (2012) Fluorescent castasterone reveals BRI1 signaling from the plasma membrane. Nat Chem Biol 8(6):583–589
6. Belkhadir Y, Yang L, Hetzel J et al (2014) The growth-defense pivot: crisis management in plants mediated by LRR-RK surface receptors. Trends Biochem Sci 39:447–456

7. Russinova E, Borst JW, Kwaaitaal M et al (2004) Heterodimerization and endocytosis of Arabidopsis brassinosteroid receptors BRI1 and AtSERK3 (BAK1). Plant Cell 16:3216–3229
8. Macho AP, Zipfel C (2014) Plant PRRs and the activation of innate immune signaling. Mol Plant 54:263–272
9. Zipfel C, Robatzek S, Navarro L et al (2004) Bacterial disease resistance in *Arabidopsis* through flagellin perception. Nature 428: 764–767
10. Sun W, Cao Y, Jansen Labby K et al (2012) Probing the *Arabidopsis* flagellin receptor: FLS2-FLS2 association and the contributions of specific domains to signaling function. Plant Cell 24:1096–1113
11. Chinchilla D, Zipfel C, Robatzek S et al (2007) A flagellin-induced complex of the receptor FLS2 and BAK1 initiates plant defence. Nature 448:497–500
12. Heese A, Hann DR, Gimenez-Ibanez S et al (2007) The receptor-like kinase SERK3/ BAK1 is a central regulator of innate immunity in plants. Proc Natl Acad Sci USA 104: 12217–12222
13. Somerville C (2006) Cellulose synthesis in higher plants. Annu Rev Cell Dev Biol 22: 53–78
14. Scheller HV, Ulvskov P (2010) Hemicelluloses. Annu Rev Plant Biol 61:263–289
15. Atmodjo MA, Hao Z, Mohnen D (2013) Evolving views of pectin biosynthesist. Annu Rev Plant Biol 64:747–779
16. McFarlane HE, Döring A, Persson S (2014) The cell biology of cellulose synthesis. Annu Rev Plant Biol 65:69–94
17. Chen L-Q, Cheung LS, Feng L et al (2015) Transport of sugars. Annu Rev Biochem 84:865–894
18. Pinto E, Ferreira IMPLVO (2015) Cation transporters/channels in plants: tools for nutrient biofortification. J Plant Physiol 179:64–82
19. Bhat RA, Panstruga R (2005) Lipid rafts in plants. Planta 223:5–19
20. Ephritikhine G, Ferro M, Rolland N (2004) Plant membrane proteomics. Plant Physiol Biochem 42:943–962
21. Morel J, Claverol S, Mongrand S et al (2006) Proteomics of plant detergent-resistant membranes. Mol Cell Proteomics 5:1396–1411
22. Griffin NM, Schnitzer JE (2011) Overcoming key technological challenges using mass spectrometry for mapping cell surfaces in tissues. Mol Cell Proteomics 10:R110.000935
23. Zhang ZJ, Peck SC (2011) Simplified enrichment of plasma membrane proteins for proteomic analyses in *Arabidopsis thaliana*. Proteomics 11:1780–1788
24. Widell S, Lundborg T, Larsson C (1982) Plasma membranes from oats prepared by partition in an aqueous polymer two-phase system. On the use of light-induced cytochrome *b* reduction as a marker for the plasma membrane. Plant Physiol 70:1429–1435
25. Albertsson P-Å, Cajarville A, Brooks DE et al (1987) Partition of proteins in aqueous polymer two-phase systems and the effect of molecular weight of the polymer. Biochim Biophys Acta 926:87–93
26. Johansson F, Olbe M, Sommarin M et al (1995) Brij 58, a polyoxyethylene acyl ether, creates membrane vesicles of uniform sidedness. A new tool to obtain inside-out (cytoplasmic side-out) plasma membrane vesicles. Plant J 7:165–173
27. Kaakinen M, Papponen H, Metsikkö K (2008) Microdomains of endoplasmic reticulum within the sarcoplasmic reticulum of skeletal myofibers. Exp Cell Res 314:237–245
28. Zhang Z, Voothuluru P, Yamaguchi M et al (2013) Developmental distribution of the plasma membrane-enriched proteome in the maize primary root growth zone. Front Plant Sci 4:33
29. Smith JM, Salamango DJ, Leslie ME et al (2014) Sensitivity of Flg22 is modulated by ligand-induced degradation and de novo synthesis of the endogenous flagellin-receptor FLAGELLIN-SENSING2. Plant Physiol 164:440–454
30. Smith JM, Leslie ME, Robinson SJ et al (2014) Loss of *Arabidopsis thaliana* dynamin-related protein 2B reveals separation of innate immune signaling pathways. PLoS Pathog 10:e1004578

Chapter 14

Probing Activation and Deactivation of the BRASSINOSTEROID INSENSITIVE1 Receptor Kinase by Immunoprecipitation

Sara Martins, Grégory Vert, and Yvon Jaillais

Abstract

Brassinosteroids (BRs) are sterol-derived hormones that control plant growth and development. The BR receptor complex is encoded by the BRASSINOSTEROID INSENSITIVE1 (BRI1) and members of the SOMATIC EMBRYOGENESIS RECEPTOR KINASE family. For activation and deactivation, the BR receptor complex uses different posttranslational modifications and recruits various partner proteins. Here, we describe optimized immunoprecipitation protocols and variants for biochemical analyses of posttranslational modifications of BRI1 and of protein–protein interactions.

Key words Brassinosteroids, BRI1, Immunoprecipitation, Ubiquitination, Phosphorylation, Protein–protein interaction

1 Introduction

Brassinosteroids (BRs) are plant steroid hormones that play important roles in plant growth and development [1–3]. Mutants defective in BR biosynthesis or signaling display severe developmental defects, such as reduced seed germination, extreme dwarfism, photomorphogenesis in the dark, altered distribution of stomata, delayed flowering, and male sterility [1, 3]. BRs are perceived at the plasma membrane by a pair of receptor kinases, including BRASSINOSTEROID INSENSITIVE1 (BRI1) and five members of the SOMATIC EMBRYOGENESIS RECEPTOR KINASE (SERK) family, with SERK3 also designated BRI1-ASSOCIATED KINASE1 (BAK1) [4–12]. The perception of BRs by the BRI1/BAK1 receptor complex triggers transphosphorylation between the BRI1 and BAK1 kinase domains in the cytoplasm [13, 14]. Downstream BR signaling involves a succession of phosphorylation and dephosphorylation events with kinases and phosphatases that culminates in the dephosphorylation of BRI1-ETHYL

Eugenia Russinova and Ana I. Caño-Delgado (eds.), *Brassinosteroids: Methods and Protocols*, Methods in Molecular Biology, vol. 1564, DOI 10.1007/978-1-4939-6813-8_14, © Springer Science+Business Media LLC 2017

METHANESULFONATE-SUPPRESSOR1 (BES1) and BRASSINAZOLE RESISTANT1 (BZR1) transcription factors and BR-responsive gene transcription [1, 15–17]. In the absence of ligand or upon deactivation of the BR receptor complex, the BRI1 kinase domain is bound by the BRI1 KINASE INHIBITOR1 (BKI1) that prevents interaction with BAK1 and spurious activation of the pathway [18, 19].

Activation or deactivation of the receptor complex relies on posttranslational modifications and interaction with positive and negative regulators. Over the past decade, extensive knowledge was gained on the phosphorylation of BRI1 [1, 13, 14, 20–24]. Several phosphorylation sites have been mapped or more recently identified by mass spectrometry after ligand binding. The timing of phosphorylation after addition of BRs together with detailed analyses of mutant variants of BRI1 impaired for specific phosphorylation sites unraveled sites playing a positive and negative role on BRI1 activity [1, 13, 14, 21–23]. In addition, BRI1 was demonstrated to be posttranslationally modified by ubiquitination and, more specifically, K63 polyubiquitin chains [25]. Although mostly independent of ligand binding to the receptor complex, BRI1 ubiquitination appears to be an important mechanism to downregulate BRI1 activity. Identification of an ubiquitination site by mass spectrometry coupled to model-based prediction allowed the generation of a non-ubiquitinatable BRI1 variant [25]. Such a ubiquitination-defective BRI1 shows a reduced internalization from the cell surface and vacuolar degradation, leading to increased plasma membrane levels and BR hypersensitivity phenotypes.

The analysis of BRI1 activation and deactivation mechanisms is mostly based on protein–protein interaction detection (such as interaction with BAK1 and BKI1) and posttranslational modifications (such as phosphorylation and ubiquitination) [13, 18, 19, 23, 25–27]. Both approaches rely on prior purification of the BRI1 protein by immunoprecipitation [18, 25, 26]. Although antibodies recognizing endogenous BRI1 proteins may be used, most studies aiming at immunoprecipitating BRI1 use anti-tag antibodies targeting functional tagged fusions of BRI1, such as, for example, functional BRI1-mCitrine [18, 25, 26], BRI1-Flag [13, 14], or BRI1-6× hemagglutinin (HA) [18]. Immunoblots using anti-tag antibodies after immunoprecipitation are used to confirm the enrichment in BRI1 compared to the input of total protein extracts. Subsequently, partner proteins or posttranslational modifications are detected with specific antibodies, such as anti-phosphoserine/threonine, anti-phosphotyrosine, anti-ubiquitin, or antibodies targeting a given protein (e.g., native or tagged versions of BAK1, BRI1, or BKI1) [13, 18, 23, 25, 26]. Subtle differences in the immunoprecipitation protocols must be considered according to the goal of the experiment. Whereas detection of reversible posttranslational modifications requires the use of specific inhibitors that prevent removal of the covalently attached modification, it also requires more stringent solubilization and wash conditions to avoid copurification of other proteins also modified and that may migrate at a molecular weight close to that of BRI1. However, for protein–protein interactions, milder conditions are used to allow detection of partner proteins. Following, we provide detailed protocols to investigate the activation and deactivation mechanisms of BRI1.

2 Materials

Prepare all the solutions with ultrapure water (resistivity of 18.2 MΩ cm at 25 °C). Keep the reagents at room temperature (unless indicated otherwise). Follow the regulation for waste disposals.

2.1 Protein Extraction from Plants and Immuno-precipitation

Keep the solutions and material at 4 °C, unless indicated otherwise.

1. Mortar and pestle at room temperature to grind tissues.
2. 1 M Tris–HCl (pH 8.0). In a glass beaker, mix approximately 50 mL of water and 12.11 g of Tris base. Once the mixture is homogenized, adjust the pH to 8.0 by adding hydrochloric acid (HCl). Make up the volume to 100 mL and store at 4 °C.
3. 0.5 M Ethylenediaminetetraacetic acid disodium salt (EDTA) (pH 8.0). In a glass beaker with 400 mL of water, mix 93.06 g of EDTA and stir with a magnetic stirrer. Adjust the pH to 8.0 with sodium hydroxide (NaOH). Make up the volume to 500 mL and autoclave the solution at 120 °C.
4. Sucrose buffer: 20 mM Tris–HCl (pH 8.0), 1 mM EDTA (pH 8.0), 0.33 M sucrose. In approximately 50 mL of water, add 2 mL of 1 M Tris–HCl (pH 8.0), 200 μL of 0.5 M EDTA (pH 8.0), and 11.3 g of sucrose. Dissolve the sucrose with the help of a magnetic stirrer and make up the volume to 100 mL with water. Autoclave the solution at 120 °C or filter sterilize. Keep at 4 °C.
5. Protease inhibitor cocktail.
6. Phosphatase inhibitor cocktail.
7. 20 mM N-ethylmaleimide (N-EM) solution in sucrose buffer: dissolve 25 mg of N-EM in 100 μL of 100 % (v/v) ethanol (*see* **Note 1**) and dilute this solution in 10 mL of sucrose buffer.
8. Sucrose buffer for ubiquitination studies: add 100 μL of protease inhibitor cocktail, and 100 μL of 20 mM N-EM solution into 10 mL of freshly prepared sucrose buffer (*see* **Note 2**).
9. Sucrose buffer for phosphorylation studies: add 100 μL of protease inhibitor cocktail, and one tablet of phosphatase inhibitor cocktail into 10 mL of freshly prepared sucrose buffer (*see* **Note 2**).

10. Miracloth.
11. Centrifuge (for instance, Sorval Legend Micro 21R, Thermo Fisher Scientific).
12. 1 M Tris–HCl (pH 6.8). Proceed as for the preparation of 1 M Tris–HCl (pH 8.0) described in **item 2**, but adjust the pH to 6.8 with HCl.
13. 4× Laemmli buffer: 2.8 mL of water, 2.4 mL of 1 M Tris–HCl (pH 6.8), 0.8 g of sodium dodecyl sulfate (SDS), 4 mL of 100 % (v/v) glycerol, 0.01 % (w/v) bromophenol blue, and 1 mL of β-mercaptoethanol (β-ME). Keep at −20 °C.
14. 2 mL Potter tissue grinder.
15. Lysis buffer: 150 mM NaCl, 1 % Triton® X-100, 50 mM Tris–HCl (pH 8.0). μMACS™ Epitope Tag Protein Isolation Kits (MACS Milteny Biotec).
16. 20 % (w/v) SDS: weigh 20 g SDS in a beaker. Add 70 mL of ultrapure water and mix gently. Adjust final volume to 100 mL with ultrapure water.
17. Rotating wheel.
18. Heated block (for instance, ThermoMixer®, Eppendorf).
19. Anti-GFP MicroBeads (*see* **Note 3**). μMACS™ Epitope Tag Protein Isolation Kits (MACS Miltenyi Biotec).
20. Anti-HA MicroBeads (*see* **Note 3**). μMACS™ Epitope Tag Protein Isolation Kits (MACS Milteny Biotec).
21. μColumns: MACS® Separation Columns (MACS Milteny Biotec).
22. MACS MultiStand (MACS Milteny Biotec).
23. Wash buffer 1: 150 mM NaCl, 1 % Igepal CA-630 (formerly NP-40), 0.5 % (w/v) sodium deoxycholate, 0.1 % (w/v) SDS, 50 mM Tris–HCl (pH 8.0). This buffer is supplied with the μMACS™ Epitope Tag Protein Isolation Kits (MACS Milteny Biotec).
24. Wash buffer 2: 20 mM Tris–HCl (pH 7.5). This buffer is provided with the μMACS™ Epitope Tag Protein Isolation Kits (MACS Milteny Biotec).
25. Elution buffer: 50 mM Tris–HCl (pH 6.8), 50 mM dithiothreitol (DTT), 1 % (w/v) SDS, 1 mM EDTA, 0.005 % (w/v) bromophenol blue, 10 % (v/v) glycerol. This buffer is provided with the μMACS™ Epitope Tag Protein Isolation Kits (MACS Milteny Biotec).

2.2 Immunoblots

1. Nitrocellulose membrane.
2. Orbital shaker.
3. Petri dishes or a small container/dish in which the nitrocellulose membrane can fit with some liquid.

4. Blotting tweezer to hold the nitrocellulose membrane.
5. 10× Tris-buffered saline (TBS): 20 mM Tris–HCl (pH 7.4), 500 mM NaCl. In approximately 800 mL of water, dissolve 24 g of Tris-base and 88 g of NaCl. Once the mixture is homogenized, adjust the pH to 7.4 by adding HCl. Make up volume to 1 L and store at room temperature.
6. Polyoxyethylene sorbitan monolaurate (Tween 20).
7. Washing solution (TBS-T): 1× TBS, 0.1 % (v/v) Tween 20 (*see* **Note 4**). Dilute 100 mL of 10× TBS in 800 mL of water. Add 1 mL of Tween 20 and adjust volume to 1 L. Store at room temperature.
8. Bovine serum albumin (BSA).
9. Blocking solution: Weight 1 g of BSA and mix gently with 90 mL of TBS-T in a beaker with the help of a stirrer (*see* **Note 5**). Adjust volume to 100 mL with TBS-T and store at 4 °C.
10. Primary antibodies: anti-green fluorescent protein-horseradish peroxidase (GFP-HRP) conjugated monoclonal antibody (MACS Miltenyi Biotec), anti-HA-HRP-conjugated monoclonal antibody (MACS Miltenyi Biotec), ubiquitin P4D1 mouse monoclonal antibody (Cell Signaling Technology), lysine63-specific anti-ubiquitin rabbit monoclonal (Apu3 clone; Merck Millipore), anti-phosphothreonine antibody (Cell Signaling Technology), anti-phosphotyrosine HRP-coupled monoclonal antibody (Millipore), primary antibody recognizing a downstream BRI1 partner or a tagged downstream partner, such as BAK1-mCherry [18, 26] or BKI1-mCitrine [18].
11. Secondary antibodies: Goat anti-mouse immunoglobulin G (heavy and light chains) (IgG [H + L])-horseradish peroxidase (HRP) conjugate and goat anti-rabbit IgG (H + L)-HRP conjugate.
12. Chemiluminescent substrate, such as Super Signal® West Dura Extended Duration Subtract (Thermo Fisher Scientific).
13. Transparent office file sheet protector.
14. Chemiluminescence imaging system.
15. 1.5-mL and 2-mL Eppendorf tubes.

3 Methods

3.1 Protein Extraction from Plants and Immunoprecipitation

Prechill the material to be used and keep everything on ice or in the cold room (*see* **Note 6**).

1. Grind tissues in liquid nitrogen (*see* **Note 7**) and transfer with a prechilled spatula to 2 mL of ice-cold sucrose buffer completed with the desired inhibitors according to the posttranslational

modification analyzed. It is also possible to use fresh tissues and grind it directly in ice-cold sucrose buffer (*see* **Note 7**).

2. Centrifuge for 10 min at 5000 × *g* at 4 °C to remove cellular debris and transfer the supernatant into 2-mL Eppendorf tubes.
3. Repeat **step 2** until the supernatant is clean (*see* **Note 8**).
4. Preheat a heat block at 95 °C.
5. Pipette 50 μL of the total protein fraction to be used as input of the immunoprecipitation. Add 15 μL of 4× Laemmli buffer.
6. Boil the input for 3 min at 95 °C. Keep the input on ice if you are planning to proceed with the Western blot on the same day or freeze in liquid nitrogen and keep at −80 °C for later use.
7. To obtain the membrane protein fraction, centrifuge for 45 min at 20,000 × *g* at 4 °C and keep the pellet (microsomal fraction) (*see* **Note 9**).
8. Add 1 mL of lysis buffer to the pellet (*see* **Note 10**) and resuspend carefully by pipetting up and down (*see* **Note 11**).
9. Transfer the lysate to ice-cold Potter tissue grinder and homogenize (*see* **Note 12**). Transfer the samples to new 1.5-mL tubes.
10. Solubilize the microsomal fraction on a rotating wheel at 4 °C for 30 min (Fig. 1).
11. Centrifuge the samples for 10 min at 20,000 × *g* at 4 °C to pellet non-solubilized microsomes. Transfer the supernatant into a new 1.5-mL tube. This fraction corresponds to solubilized microsomal proteins and will be used for BRI1 immunoprecipitation.
12. Quantify the samples and use 1 mg of proteins of the supernatant to perform the immunoprecipitation of BRI1.
13. Add 50 μL of anti-tag MicroBeads to the supernatant and mix well (Fig. 1).
14. Incubate on ice for 30 min to allow the formation of the BRI1-tag-antibody complex.
15. Place the μColumn on the μMACS Separator and apply 200 μL of the lysis buffer to equilibrate the column (*see* **Note 10**).
16. Pipette 75 μL of elution buffer for each sample into a new 1.5-mL tube and boil it on the preheated block at 95 °C.
17. Apply the sample lysate onto the center of the column and let it run through by gravity flow.
18. Wash the column with 200 μL of wash buffer 1. Repeat four times.
19. Wash the column with 100 μL of wash buffer 2.

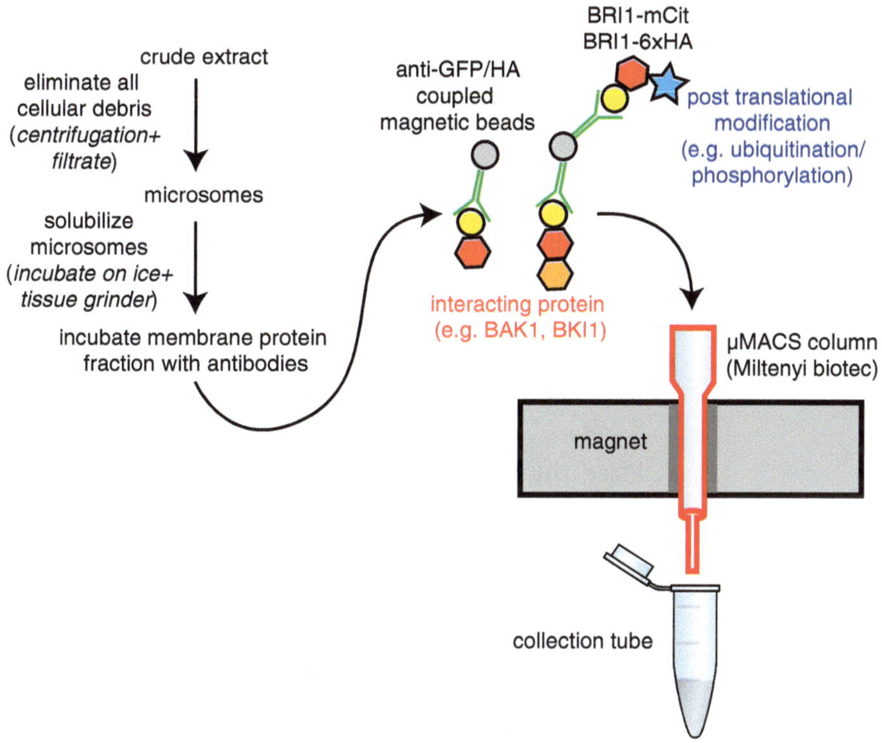

Fig. 1 Protein extraction and purification scheme described in this protocol

20. Apply 20 μL of elution buffer kept at 95 °C onto the center of the column and incubate at room temperature for 5 min (*see* **Note 13**).
21. Place a new 1.5-mL tube under the column.
22. Apply 50 μL of elution buffer kept at 95 °C onto the center of the column and collect the eluates (*see* **Note 14**). Freeze samples in liquid nitrogen and keep them at −80 °C, or proceed directly with immunoblotting.

3.2 Immunoblots

Run the input and immunoprecipitated proteins in a standard SDS-polyacrylamide gel electrophoresis with a 7.5 % separating gel (*see* **Note 15**). Transfer the proteins from the gel to a nitrocellulose membrane (*see* **Note 16**). It is important to handle the membrane with blotting tweezers and to keep it wet at all times. All procedures can be done at room temperature unless specified otherwise.

3.2.1 Detection of BRI1-mCitrine or BRI1-6×HA by Western Blot

1. Place the nitrocellulose membrane in a Petri dish and add the blocking solution until it covers the membrane. Shake gently for 1 h at room temperature on the orbital shaker.
2. Prepare the solution containing the primary antibodies by adding anti-GFP-HRP or anti-HA-HRP antibodies in a

new blocking solution. Use 1 μL of antibody to 5 mL of blocking solution.

3. Replace the blocking solution with the antibody solution. Incubate for 1 h shaking gently on the orbital shaker (*see* **Note 17**).
4. Discard the antibody solution.
5. Wash the membrane three times for 10 min in washing solution under gentle agitation.
6. Discard the washing solution.
7. Apply the Super Signal® West Dura Extended Duration Subtract as indicated by the manufacturer and incubate for 5 min.
8. Place the membrane between a transparent office file sheet protector.
9. Proceed to signal detection with a chemiluminescence imaging system.

3.2.2 Detection of BRI1 Ubiquitination, Phosphorylation, or Coimmunoprecipitated Proteins

1. Place the nitrocellulose membrane in a Petri dish and add the blocking solution until it covers the membrane and shake gently during 1 h at room temperature on the orbital shaker (*see* **Note 17**).
2. Prepare the solution with the primary antibody in a new blocking solution. Use the following dilutions: anti-P4D1 1:2500; anti-Apu3 1:2000; anti-pThr 1:1000; anti-pTyr 1:5000. *See* Fig. 2 for an example of Western blot with anti-GFP and antiubiquitin antibodies.
3. Discard the blocking solution and add the antibody solution. Incubate for 1 h and shake gently on the orbital shaker.
4. Discard the antibody solution.

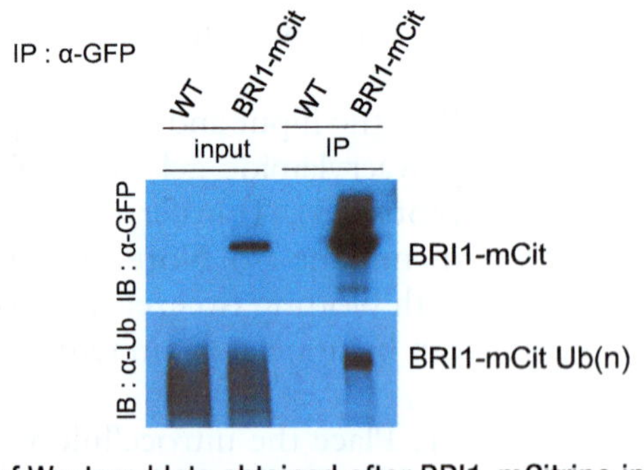

Fig. 2 Example of Western blots obtained after BRI1-mCitrine immunoprecipitation (IP). IP was done with anti-GFP antibodies on solubilized protein extracts from wild-type (WT) and transgenic plants expressing BRI1-mCitrine (BRI1-mCit) and subjected to immunoblotting (IB) with anti-GFP and antiubiquitin (Ub) antibodies

5. Wash the membrane three times for 10 min in washing solution under gentle agitation.
6. Discard the washing solution.
7. Prepare the HRP-conjugated secondary antibody in blocking solution (dilution 1:20,000).
8. Wash the membrane three times for 10 min in the washing solution with gentle shaking.
9. Discard the washing solution.
10. Apply the Super Signal® West Dura Extended Duration Subtract according to the manufacturer's instruction and incubate for 5 min.
11. Place the membrane between a transparent office file sheet protector.
12. Proceed with signal detection with a chemiluminescence imaging system.

4 Notes

1. As the N-EM does not dissolve in the sucrose buffer, it is necessary to first dissolve it in 100 % (v/v) ethanol.
2. N-EM is an efficient deubiquitinating enzyme inhibitor that prevents the potential deubiquitination of ubiquitinated proteins, including BRI1. Phosphatase inhibitors will prevent dephosphorylation of phosphorylated proteins. It is strongly advised to add the protease inhibitor cocktail, phosphatase inhibitor cocktail, and N-EM extratemporaneously to the sucrose buffer just before protein extraction and immunoprecipitation. The protease inhibitor cocktail is important to avoid protein degradation.
3. To perform the immunoprecipitation, we routinely use the tagged forms of BRI1 such as BRI1-6×HA [18] or BRI1-mCitrine [18, 25, 26] that can be recognized by the anti-HA or anti-GFP MicroBeads used in this protocol, respectively.
4. The TBS is recommended for proteins that are phosphorylated, because it prevents a high background on the nitrocellulose membrane during the Western blot washes. Tween 20 is a detergent that will help avoid non-specific binding to the membrane.
5. The bovine protein can be degraded at high temperatures mechanically; therefore, it should be stored at 4 °C. BSA is important for blocking the membrane prior and during the incubation with the antibodies specific for the protein of interest in order to avoid non-specific binding of the antibody to the membrane.

6. Work on ice and/or use a cold room (4 °C). It is important that everything remains cold at all times.
7. Keep the mortar and pestle at room temperature to avoid water condensation and the formation of ice crystals when in contact with the liquid nitrogen. Cool down the mortar and pestle with liquid nitrogen. Leave a small quantity of liquid nitrogen inside the mortar to help keep the tissue frozen while grinding. It is recommended to start with at least 1 g of tissue. Tissues must be kept in liquid nitrogen or at −80 °C before being ground. It is also possible to grind fresh tissue directly in the ice-cold sucrose buffer. This latter option is preferable when the plant material (for instance, seedlings) has been submitted to liquid treatments prior to extraction in order to avoid the formation of ice following incubation in liquid nitrogen.
8. These steps are critical because it is important to have a clean supernatant to prevent that cellular debris are carried to the membrane fraction. Usually, it takes five extra centrifugations at 4 °C at 5000 × *g* for 10 min to obtain a clean supernatant. It is advisable to stay well clear of the pellet when pipetting the supernatant. Before the initial first two centrifugations, it is possible to quickly filter out the bulk of the cellular debris by passing the solution through Miracloth. It is important to always use an ice-cold buffer and to work at 4 °C or on ice.
9. The supernatant resulting from the 20,000 × *g* centrifugation corresponds to the soluble protein-enriched fraction and the pellet contains the microsomal fraction, in which BRI1 is enriched.
10. The analysis of posttranslational modifications of BRI1 requires more stringent conditions to strip off interacting proteins by adding 0.1 % (w/v) SDS to the lysis buffer. In that case, the μMACS column must be equilibrated with the SDS-containing lysis buffer.
11. If the same sample is separated in two 2-mL tubes, pipette 500 μL of lysis buffer into each tube. Resuspend the pellet carefully by pipetting up and down, and avoid pipetting the foam.
12. To avoid foaming while using the potter tissue grinder, we recommend to keep it submerged in the sample.
13. The μColumn has a capacity of 20 μL. When the first 20 μL of elution buffer is added, it is important to make sure that wash buffer 2 runs completely through the column to avoid residual salts and detergents from this wash buffer on the eluate sample.
14. It is possible to improve the eluate sample collection by placing Parafilm (or a pipette rubber bulb) on top of the column and pushing it to create a little pressure allowing the eluate that is retained in the column to come out.

15. BRI1 is a 130-kDa protein that reaches 160 kDa when coupled with GFP. For this reason, we use 7.5 % polyacrylamide gels, but when low-molecular-weight interacting proteins are expected, it is also possible to use gels with a higher polyacrylamide concentration.
16. In the case of BRI1, we recommend to use 20 % (v/v) methanol during the transfer.
17. It is important that the membrane is always in contact with the solution during the antibody incubation time. Hence, the agitation speed of the orbital shaker can be optimized.

Acknowledgments

We thank Youssef Belkhadir for providing tips during the establishment of this protocol. Y.J. has received funding from the European Research Council (3363360-APPL) and from the Marie Curie Action (PCIG-GA-2011-303601) under the European Union's Seventh Framework Programme (FP/2007-2013). G.V. has received funding from Agence Nationale de la Recherche (ANR-13-JSV2-0004-01), and from the Marie Curie Action (PCIG12-GA-2012-334021 under the European Union's Seventh Framework Programme (FP/2007-2013). S.M. is sponsored by a PhD fellowship from the Saclay Plant Sciences LabEx initiative (ANR-10-LABX- 0040-SPS) funded by the French government.

References

1. Belkhadir Y, Jaillais Y (2015) The molecular circuitry of brassinosteroid signaling. New Phytol 206:522–540
2. Jaillais Y, Chory J (2010) Unraveling the paradoxes of plant hormone signaling integration. Nat Struct Mol Biol 17:642–645
3. Vert G, Nemhauser JL, Geldner N et al (2005) Molecular mechanisms of steroid hormone signaling in plants. Annu Rev Cell Dev Biol 21:177–201
4. Nam KH, Li J (2002) BRI1/BAK1, a receptor kinase pair mediating brassinosteroid signaling. Cell 110:203–212
5. Li J, Chory J (1997) A putative leucine-rich repeat receptor kinase involved in brassinosteroid signal transduction. Cell 90:929–938
6. Li J, Wen J, Lease KA et al (2002) BAK1, an *Arabidopsis* LRR receptor-like protein kinase, interacts with BRI1 and modulates brassinosteroid signaling. Cell 110:213–222
7. Wang Z-Y, Seto H, Fujioka S et al (2001) BRI1 is a critical component of a plasma-membrane receptor for plant steroids. Nature 410:380–383
8. He Z, Wang Z-Y, Li J et al (2000) Perception of brassinosteroids by the extracellular domain of the receptor kinase BRI1. Science 288:2360–2363
9. Albrecht C, Russinova E, Kemmerling B et al (2008) Arabidopsis SOMATIC EMBRYOGENESIS RECEPTOR KINASE proteins serve brassinosteroid-dependent and -independent signaling pathways. Plant Physiol 148:611–619
10. Gou X, Yin H, He K et al (2012) Genetic evidence for an indispensable role of somatic embryogenesis receptor kinases in brassinosteroid signaling. PLoS Genet 8:e1002452
11. Santiago J, Henzler C, Hothorn M (2013) Molecular mechanism for plant steroid receptor activation by somatic embryogenesis co-receptor kinases. Science 341:889–892
12. Hothorn M, Belkhadir Y, Dreux M et al (2011) Structural basis of steroid hormone perception by the receptor kinase BRI1. Nature 474:467–471

13. Wang X, Kota U, He K et al (2008) Sequential transphosphorylation of the BRI1/BAK1 receptor kinase complex impacts early events in brassinosteroid signaling. Dev Cell 15:220–235
14. Wang X, Goshe MB, Soderblom EJ et al (2005) Identification and functional analysis of in vivo phosphorylation sites of the Arabidopsis BRASSINOSTEROID-INSENSITIVE1 receptor kinase. Plant Cell 17:1685–1703
15. Vert G, Chory J (2006) Downstream nuclear events in brassinosteroid signalling. Nature 441:96–100
16. Tang W, Yuan M, Wang R et al (2011) PP2A activates brassinosteroid-responsive gene expression and plant growth by dephosphorylating BZR1. Nat Cell Biol 13:124–131
17. Kim T-W, Guan S, Sun Y et al (2009) Brassinosteroid signal transduction from cell-surface receptor kinases to nuclear transcription factors. Nat Cell Biol 11:1254–1260
18. Jaillais Y, Hothorn M, Belkhadir Y et al (2011) Tyrosine phosphorylation controls brassinosteroid receptor activation by triggering membrane release of its kinase inhibitor. Genes Dev 25:232–237
19. Wang X, Chory J (2006) Brassinosteroids regulate dissociation of BKI1, a negative regulator of BRI1 signaling, from the plasma membrane. Science 313:1118–1122
20. Wu X, Oh M-H, Kim HS et al (2012) Transphosphorylation of *E. coli* proteins during production of recombinant protein kinases provides a robust system to characterize kinase specificity. Front Plant Sci 3:262
21. Oh M-H, Wang X, Clouse SD et al (2012) Deactivation of the *Arabidopsis* BRASSINOSTEROID INSENSITIVE 1 (BRI1) receptor kinase by autophosphorylation within the glycine-rich loop. Proc Natl Acad Sci USA 109:327–332
22. Oh M-H, Clouse SD, Huber SC (2012) Tyrosine phosphorylation of the BRI1 receptor kinase occurs via a post-translational modification and is activated by the juxtamembrane domain. Front Plant Sci 3:175
23. Oh M-H, Wang X, Kota U et al (2009) Tyrosine phosphorylation of the BRI1 receptor kinase emerges as a component of brassinosteroid signaling in *Arabidopsis*. Proc Natl Acad Sci USA 106:658–663
24. Bojar D, Martinez J, Santiago J et al (2014) Crystal structures of the phosphorylated BRI1 kinase domain and implications for brassinosteroid signal initiation. Plant J 78:31–43
25. Martins S, Dohmann EMN, Cayrel A et al (2015) Internalization and vacuolar targeting of the brassinosteroid hormone receptor BRI1 are regulated by ubiquitination. Nat Commun 6:6151
26. Jaillais Y, Belkhadir Y, Balsemão-Pires E et al (2011) Extracellular leucine-rich repeats as a platform for receptor/coreceptor complex formation. Proc Natl Acad Sci USA 108:8503–8507
27. Fàbregas N, Li N, Boeren S et al (2013) The BRASSINOSTEROID INSENSITIVE1-LIKE3 signalosome complex regulates *Arabidopsis* root development. Plant Cell 25:3377–3388

Chapter 15

The Primary Root of *Sorghum bicolor* (L. Moench) as a Model System to Study Brassinosteroid Signaling in Crops

David Blasco-Escámez*, Fidel Lozano-Elena*, Norma Fàbregas, and Ana I. Caño-Delgado

Abstract

Roots anchor plants to the soil and are essential for a successful plant growth and adaptation to the environment. Research on the primary root in the plant model system *Arabidopsis thaliana* has yielded important advances in the molecular and cellular understanding of root growth and development. Several studies have uncovered how the hormones brassinosteroids (BRs) control cell cycle and differentiation programs through different cell-specific signaling pathways that are key for root growth and development. Currently, an important challenge resides in the translation of the current knowledge on *Arabidopsis* roots into agronomically valuable species to improve the agricultural production and to meet the food security goals of the millennium. In this chapter, we characterize the primary root apex of the cereal *Sorghum bicolor* (L. Moench) (sorghum), analyze the physiological response of sorghum roots to BRs, and examine the phylogeny of the BRASSINOSTEROID INSENSITIVE1-like receptor family in *Arabidopsis* and its orthologous genes in sorghum. Overall, we support the use of sorghum as a suitable crop model species for the study of BR signaling in root growth and development. The methods presented enable any laboratory worldwide to use sorghum primary roots as a favorite organ for the study of growth and development in crops.

Key words Sorghum, Root development, Hormone, Brassinosteroid, BRI1-like receptors

1 Introduction

Sorghum bicolor (L. Moench) (sorghum) is a cereal grass (Poaceae) that is cultivated worldwide and is mainly used for animal feeding but represents only approximately 2.4 % of world's total cereal production (http://faostat.fao.org). Here, we propose sorghum as an excellent crop model for the study of root growth and development because it holds several genetic and physiological advantages. Whereas its physiological and anatomical features are similar to those of other crops with a high economic impact, such as *Zea mays* (maize) or *Oryza sativa* (rice) [1, 2], sorghum has a diploid and relatively small genome (approximately 730 Mb) [3, 4] compared to the

Eugenia Russinova and Ana I. Caño-Delgado (eds.), *Brassinosteroids: Methods and Protocols*, Methods in Molecular Biology, vol. 1564, DOI 10.1007/978-1-4939-6813-8_15,

tetraploid genome of maize [5]. In addition, sorghum is susceptible to *Agrobacterium tumefaciens*-mediated plant transformation, thus responsive to genetic engineering tools, such as Cas9-based technologies [6]. Moreover, the use of Targeted Induced Local Lesions in Genomes (TILLING) for the screening of mutants has been successfully applied and optimized in sorghum plants [7–9].

In addition to these advantages, one important feature that makes sorghum attractive from an agricultural point of view is its high drought tolerance when compared to other cereal species. Sorghum is often cultivated instead of maize in dry, populated areas of the planet. This trait is of particular interest in the context of the current climate change that calls for the production of drought-resistant crops worldwide, specially in subtropical areas [10]. The abiotic stress tolerance has been extensively studied at the molecular and physiological levels. Sorghum-specific drought tolerance traits have been described as resulting from increased cuticular wax levels that prevent further transpiration upon stomata closure [11, 12], decreased levels of oxidative stress under severe drought [13], and postflowering resistance of some varieties (such as the "stay green" trait) that involves leaf senescence retardation to keep the photosynthetic activity for a proper grain development [14]. Furthermore, quantitative trait locus (QTL) analysis has identified genomic regions associated with preflowering drought resistance and stay-green traits [15–18].

The root system architecture of sorghum is similar to that of most cereals and consist of the embryonic primary root, adventitious roots arising from the first mesocotyl node and brace roots that develop from the second node and above [2, 19]. Although brace roots play a clear role in water and nutrient uptake and resistance to lodging [2], the primary root is sufficient to support proper growth and development of the plant [20]. The sorghum roots have been linked to drought resistance and stay-green traits [21] and QTLs controlling root growth and adaptation to drought have been extensively screened in sorghum [22, 23]. However, most research efforts have focused on improving agronomic traits through breeding and very little has been done toward the genetic engineering of sorghum roots. Future research in this direction will be essential for a qualitative jump in the development of novel and better-adapted varieties.

Regarding root research, the primary root of *Arabidopsis* is the most widely investigated model for root growth and development, but specifically in the case of cereals, maize roots have been the model of choice because, like sorghum, maize fundamentally maintains the spatial organization of the primary root present in *Arabidopsis* [2].

Our laboratory has extensively investigated brassinosteroid (BR) hormones as one of the major players controlling *Arabidopsis* root development processes, such as vascular patterning, root meristem maintenance, and quiescence control in the root stem cell niche

[24–26]. BRs are perceived by three plasma membrane protein receptors, specifically BRASSINOSTEROID-INSENSITIVE1 (BRI1) that is widely expressed in the primary root and two paralogs BRI1-like1 (BRL1) and BRI1-like3 (BRL3) that are specifically expressed in the quiescent center and vascular tissues [24]. Whereas BRL2 does clearly not bind the BRs [27], its potential role in BR signaling remains unknown. Recently, most of the BR signaling and biosynthetic genes have been identified in the sorghum genome and Single Nucleotide Polymorphisms (SNPs) in some of these genes have been associated with plant architecture phenotypes [28]. The BR receptors of sorghum share structures similar to the BR proteins previously reported in *Arabidopsis*, with sequence homologies ranging between 56 % and 60 % of identity. Interestingly, sorghum conserves an ortholog of the *Arabidopsis* receptor BRL2 (that does not bind BRs), but has only one single common ortholog for BRL1 and BRL3 receptors of *Arabidopsis* (Fig. 1). Conversely, in the rice and maize genomes, the same BR receptor distribution present in *Arabidopsis* is conserved; maize even shows an additional close homolog to BRI1 [29].

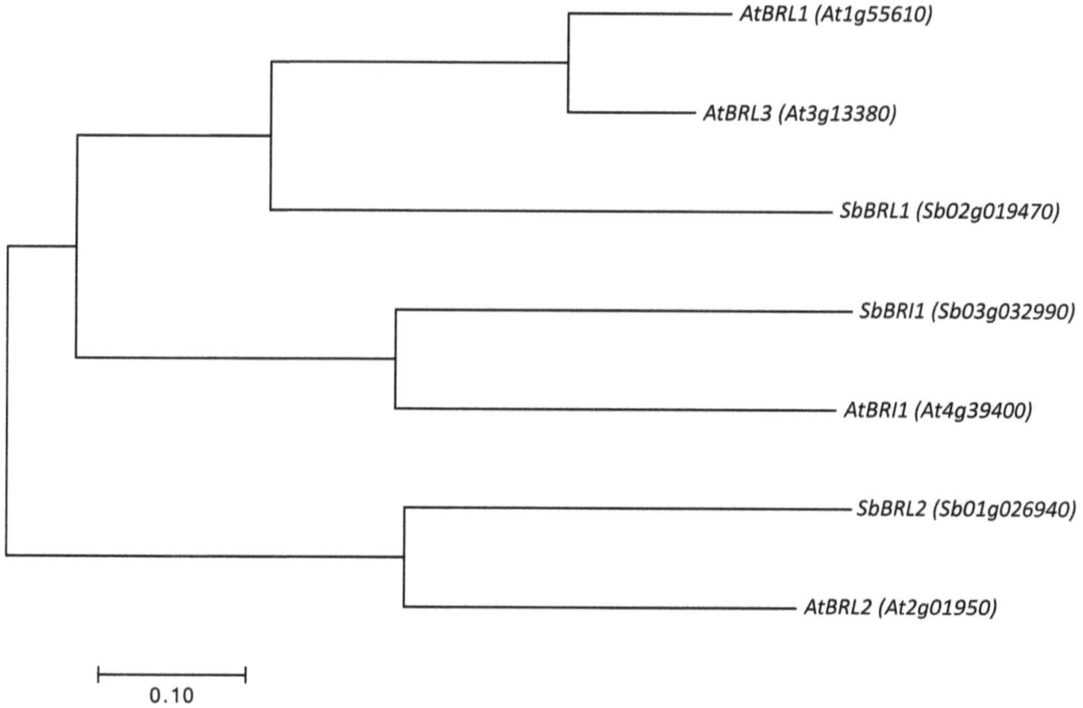

Fig. 1 Phylogenetic tree of *Sorghum bicolor* and *Arabidopsis thaliana* brassinosteroid receptors. The tree was constructed with the Maximum Likelihood method. The sequences were retrieved by BLAST search of *Arabidopsis* receptors against the sorghum draft genome [3]. The alignment was done with Clustal Omega and the tree was constructed with MEGA software (v.7.0.7). The scale represents the number of amino acid changes per site. Note that sorghum has an ortholog for the nonfunctional AtBRL2 and one common ortholog for the AtBRL1 and AtBRL3 receptors.

Given the high degree of conservation of the BR receptors in different plant species, we describe a set of experimental methods to translate our knowledge from the model plant *Arabidopsis* to the crop cereal sorghum and to use it as a crop model due to the advantages mentioned earlier and specially for its potential to become a vanguard cereal facing agricultural challenges in the near future.

In this chapter, we provide a detailed methodology to investigate the role of BRs in the sorghum primary root. In brief, sorghum seedlings can be properly grown in plates and their physiological responses to BRs are similar to those in *Arabidopsis* [24, 25]. Furthermore, we describe a method for optimal microscopy imaging of the root meristem organization and the implemented microscopy techniques to analyze the vascular pattern of sorghum roots.

2 Materials

2.1 Plant Growth and Physiological Treatments with Brassinolide

1. *Sorghum bicolor* (L. Moench) seeds.
2. Vertical plate rack.
3. Tweezers.
4. Sterile distilled water.
5. Ethanol 70 % (v/v).
6. Sterilization solution: 35 % (v/v) sodium hypochlorite solution (NaClO 15 % [w/v]) in milliQ sterile water.
7. Brassinolide (BL) ((2α, 3α, 5α, 22R, 23R, 24S)-2,3,22,23-tetrahydroxy-β-homo-7- oxaergostane-6-one) (*see* **Note 1**).
8. Dimethyl sulfoxide (DMSO) (*see* **Note 1).**
9. Water bath.
10. Square Petri dishes (12×12 cm).
11. Micropore surgical tape.
12. Growth medium: 1 % (w/v) agar-Murashige and Skoog (MS) salts supplemented with vitamins, 10 g/L of sucrose, and 5 mM 2-(*N*-morpholino)ethanesulfonic acid (MES)/KOH, pH 5.7 (*see* **Note 2**).
13. Laminar flow.
14. Steel razor blades.
15. Propidium iodide (PI) solution: 10 μg/mL in milliQ water.
16. Sterile 50 mL tubes.

2.2 Histological Embedding

1. Steel razor blades.
2. Glass scintillation vials (for instance, 20 mL) with screw caps.
3. 1.25 % (v/v) glutaraldehyde in 0.1 M sodium cacodylate buffer, pH 7.4 (electron microscopy grade).

4. Ethanol dilution series (30 %, 50 %, 70 %, 90 %, and 100 %).
5. Resin (Technovit© 7100) (*see* **Note 3**).
6. Expanded polystyrene box filled with ice.
7. Polyethylene embedding molds (for instance, Leica 8 × 7 mm).
8. Tweezers.
9. Parafilm.
10. Microtome.
11. Stylet.
12. Brush.
13. Microscope slides (76 × 26 mm).
14. Steel razor blades.
15. Glass knife razor blades (*see* **Note 4**).
16. Glass Knifemaker (such as EM KMR3 from Leica) (*see* **Note 4**).
17. Hot plate.
18. Staining solution: 0.1 % (w/v) toluidine blue in 0.1 M $NaHPO_3$, pH 7.0.
19. Glass staining jars.
20. Water.
21. Cover slips.
22. Ethanol 70 % (v/v).
23. Axiophot microscope (for instance Zeiss).
24. Orbital shaker.

2.3 Pseudo-Schiff Reagent Adapted Protocol to Sorghum Roots

1. Fixative solution: 70 % (v/v) methanol + 10 % (v/v) acetic acid (glacial) in distilled water.
2. Distilled water.
3. 6-well cell culture plate (such as Costar from Corning).
4. Filters (for instance, cell strainer 40 μm from Falcon).
5. Pasteur pipettes.
6. Periodic acid 1 % (v/v).
7. Schiff's reagent: 0.1 M sodium metabisulfite, 0.15 N HCl in milliQ water.
8. 1 mg/mL PI in milliQ water.
9. Hoyer's solution: 30 g gum arabic, 200 g chloral hydrate, 20 g glycerol, and 50 mL distilled water [30] (*see* **Note 5**).
10. Magnetic stirrer.
11. Small brush.
12. Microscope slides (76 × 26 mm).
13. Pencil.

14. Tweezers.
15. Confocal microscope (such as FV 1000 confocal microscope from Olympus).

3 Methods

3.1 Brassinosteroid Dose Response Curve Assays

1. Heat until melt one bottle of growth medium, 1 % (w/v) agar half-strength (½) MS without sucrose and supplemented with vitamins.
2. Place the bottle with the agar medium in a water bath at 50 °C (*see* **Note 1**).
3. Calculate the amount of brassinolide (BL) needed to prepare 50 mL of growth medium supplemented with 400 nM BL.
4. From this point, always work under sterile conditions (i.e., in a laminar flow hood) and with sterile material. Aliquot 50 mL of ½ MS temperate medium and add the required amount of BL. Mix well. As BL is very sensitive to high temperatures, make sure the temperature of the medium is not higher than 50 °C to ensure an optimal BL activity in the assay.
5. Make serial BL dilutions (1:10) by taking 5 mL of the medium with the highest BL concentration and by adding 45 mL of ½ MS medium to get the succeeding dilution.
6. Pour 45 mL of the mixture into square plates.

3.2 Sorghum Seed Surface Sterilization and Sowing for Physiological Assays

1. Soak the seeds for 1 minute in 70 % (v/v) ethanol.
2. Remove ethanol and add 35 % (v/v) sodium hypochlorite solution. Stir for 30 minutes.
3. Remove the sodium hypochlorite solution and rinse the seeds with sterile distilled water 5 times for 1 minute each.
4. Soak the seeds for 1 hour with sterile distilled water.
5. Rinse the seeds with sterile distilled water 5 times for 1 minute each.
6. Sow the seeds on agar plates with the help of sterile tweezers. Place the seeds with the embryo facing downwards (*see* **Note 6**).
7. After 4 days of growth in a chamber at 28 °C and 12-hour light/12-hour dark photoperiod, photograph the plates including an appropriate scale (Fig. 2c).
8. Measure the root length of the seedlings (*see* **Note** 7).

3.3 Resin Embedding of Sorghum for Microtome Semithin Sectioning

1. Cut sections of approximately 1 cm of primary root (1 cm above the root tip). Use a steel razor blade, directly over the plate with growing medium. Proceed rapidly to the next step.

2. Transfer carefully the sections to scintillation vials with the help of tweezers. Fix the sorghum root sections in 1 % (v/v) glutaraldehyde in 0.1 M sodium cacodylate buffer, pH 7.4 (electron microscopy grade). Infiltrate the samples under vacuum for 30 minutes. Keep them at 4 °C overnight. Samples can be kept for several months at this stage.
3. Remove the fixation solution and dehydrate the samples through an ethanol wash series of 30 % (v/v), 50 %, 70 %, 90 %, and 100 % for 60 minutes each and gentle shaking in an orbital shaker at room temperature.
4. Repeat the 100 % (v/v) ethanol step to completely dehydrate the sample.
5. Remove the 100 % (v/v) ethanol and start the historesin infiltration by adding historesin-I:ethanol solution (1:1). Incubate for 60 min at room temperature with gentle shaking (*see* **Note 3**).
6. Incubate the samples with historesin-I 100 % for 60 min at room temperature with gentle agitation, then replace it with fresh historesin-I 100 % solution and repeat the 60-min incubation step.
7. Add freshly prepared historesin-I 100 % solution and keep samples at 4 °C overnight.
8. Label plastic molds and place them on ice in horizontal position (*see* **Note 8**).
9. Place root sections into each plastic mold and fill it with 100 % historesin-II. Several root sections can be placed in the same mold.
10. Distribute and orientate the samples with tweezers.
11. Cover with parafilm and kept overnight at 4 °C to activate the historesin polymerization (*see* **Note 9**).
12. Sample blocks can be stored at room temperature for months until microtome sectioning.

3.4 Microtome Sectioning, Toluidine Blue Staining, and Microscopy Imaging

1. Remove solidified blocks from the plastic molds with a stylet as prying tool.
2. Trim away excess of historesin by cutting off the corners of the block and place the block on the microtome holder for sectioning.
3. Place a glass knife in the support and cut historesin away (5–10 μm) until reaching the sample (*see* **Note 10**).
4. Cut semithin sections (5–8 μm) using a glass knife in a microtome (*see* **Note 11**).
5. Label each microscope slide with a pencil and cover it with milliQ water.

6. Collect several tissue sections with a brush and place them on the slide.
7. Evaporate water from the samples by placing the slide on a hot plate for 10 min at 80 °C.
8. Stain sections by submerging slides with 0.1 % (w/v) toluidine blue in 0.1 M $NaPO_4$ (pH 7.2) for 10 s.
9. Rinse slides with sterilized water and clean overstaining in the back of the slide with EtOH 70 % (w/v) for optimal imaging.
10. Add a drop of water and a coverslip.
11. Image with an optical microscope under white light. Set the zoom at 20× for a complete section image and at 40× for a more detailed cell-type imaging (Fig. 2f).

3.5 Pseudo-Schiff Propidium Iodide (mPS-PI) Staining for Visualization of Sorghum Roots Under Confocal Microscopy

1. Fill the 6-well plates with the fixative solution and submerge the filters.
2. Cut approximately 1 cm of root tips of 4-day-old seedlings directly over the plate. Carefully transfer them to the fixative solution-containing wells Incubate overnight at 4 °C.
3. Wash the samples twice with milliQ water in new 6-well plates and transfer the filters with the samples.
4. Transfer the samples to a 1 % (v/v) periodic acid solution for 30 min with gentle shaking at room temperature.
5. Prepare a fresh mixture of Schiff's Reagent and Propidium Iodide (1 mg/mL) in a 15:1 (w/w) proportion. Transfer the samples and incubate for 2 h at room temperature.
6. Put one drop of Hoyer's solution on a microscope slide with a Pasteur pipette, spread the solution with a brush, and transfer the samples individually with tweezers.
7. Extend the roots with the help of a small brush and carefully put a cover slip over the samples.
8. Allow the samples to dry in the dark for at least 2 days.
9. Visualize with a confocal microscopy with a diode laser line excitation (Fig. 2d; *see* **Note 12**).

4 Notes

1. BL must be dissolved in DMSO and added to the growth medium below 50 °C in order not to degrade the molecule. Storage of plates should be done at 4 °C in the dark.
2. Only after the pH of the MS medium has been adjusted to pH 5.7 with KOH, add plant agar and autoclave the bottle medium. The sorghum seedlings are grown vertically for 4 days at 28 °C with a 12-h light/12-h dark photoperiod.

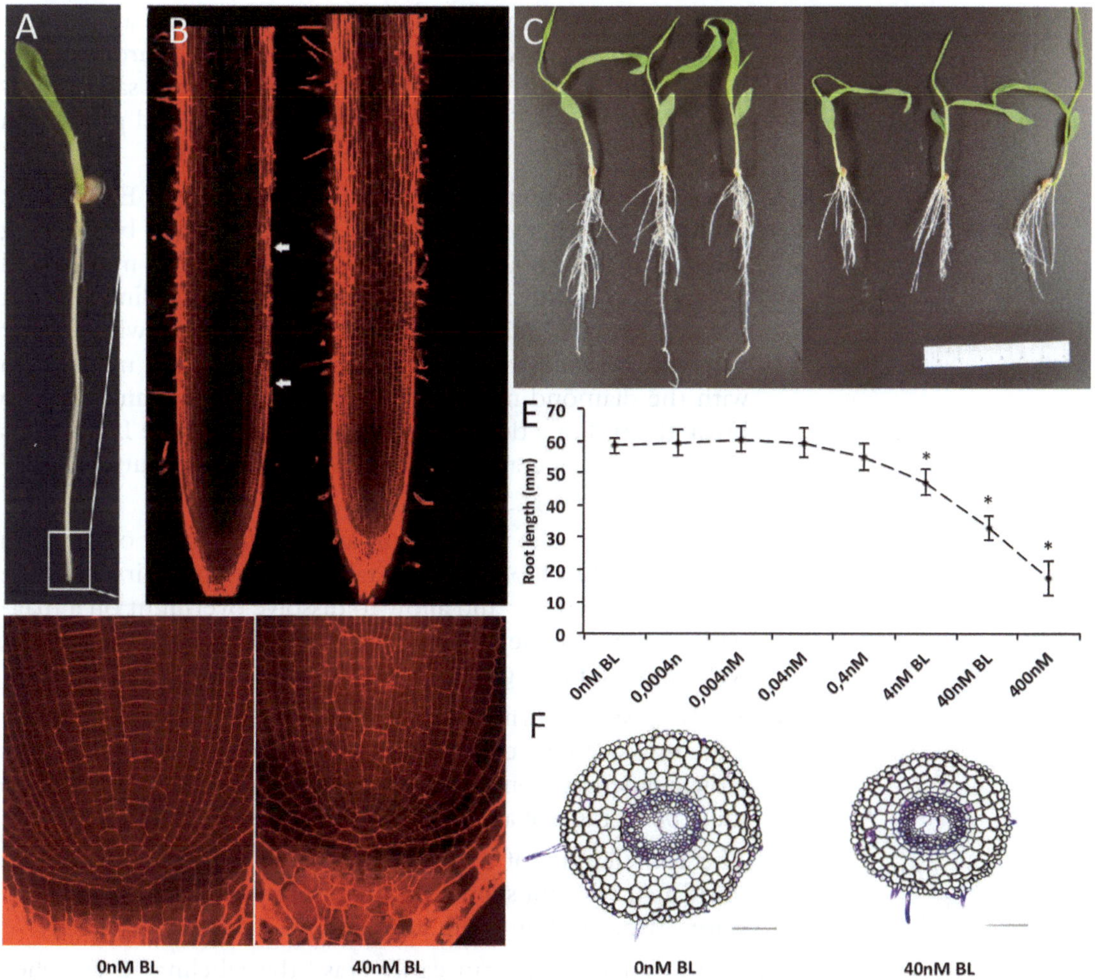

Fig. 2 Microscopic analysis of *Sorghum bicolor* primary root in response to BL treatments. (**a**) Four-day-old *Sorghum bicolor* seedling. (**b**) Detailed zoom (20×) showing root tip. Two confocal microscopy planes (central plane to the *left* and epidermis to the *right*) of in vivo PI-stained roots. *White arrows* mark the meristematic zone boundary (*bottom*) and the elongation zone boundary (*top*). (**c**) Four-day-old seedlings without treatment (control) (*left*) and exogenously treated with 40 nM of BL (*right*). (**d**) Detail of quiescent center (QC) of the primary root after the mPS-PI protocol. *Left*, control; *right*, after treatment with 40 nM BL. Note that the cell walls in the QC cells are divided after BL treatment as reported in *Arabidopsis*. (**e**) Dose–response curve of 4-day-old seedlings upon BL application. *Bars* and *asterisks* denote standard errors and significant differences in a two-tailed Student's *t*-test of three independent biological replicates (P value < 0.05), respectively. (**f**) Cross sections (5 μm) as observed under the optical microscope. The 4-day-old samples were collected in the differentiation zone of the primary root and were embedded, microtome cut, and stained with toluidine blue. Xylem and phloem bundles are stained purple and blue, respectively. Note the smaller diameter and the more developed xylem of the BL-treated root. Scale bar, 50 μm

3. The commercial kit Technovit 7100 consists of historesin solution, hardener 1, and hardener 2. To prepare the historesin-1, mix 100 mL of historesin solution with 1 g of hardener 1. For the preparation of the historesin 2, mix 15 mL of historesin 1 with 1 mL of hardener 2. Historesin 2I polymerizes fast, hence

prepare small volumes to use within a few minutes and, if necessary, prepare more again. As historesin 1 and 2 are irritating to eyes and skin, it is mandatory to take proper safety measures, namely, wear of laboratory coat, gloves, and protection glasses.

4. Glass knifes are prepared from a glass bar with an EM KMR2 device (Leica Microsystems). The procedure starts by marking the glass bar with a diamond glass marker in the middle and afterward putting pressure in the center of the line. Perfect squares can be obtained by repeating this process with smaller bar pieces. Once the glass squares are obtained, mark them with the diamond marker from corner to corner and press to break them. For a detailed video description of the Leica Glass Knifemaker operation, see https://www.youtube.com/watch?v=nXF1Mje_fNU.
5. Hoyer's solution is very viscous and solubilization of the gum arabic with the chloral hydrate takes a long time. First, add the gum arabic to water and allow to dissolve overnight on a magnetic stirrer. The next day, slowly add chloral hydrate and mix well before adding glycerol.
6. Seeds must be sown inside agar to prevent them from falling. Take into account that plates will be placed vertically and the radicle emerges from the embryo (dark spot). Hence, embryo must be facing the agar and oriented downward.
7. The root length of 4-day-old seedlings grown vertically on ½MS agar without sucrose was measured with the ImageJ software (http://rsb.info.nih.gov/ij/).
8. Historesin solution can easily erase the labeling, even when done with a permanent marker.
9. Make sure to cover the sample without bubbles or air inside the block, because it will not permit a homogeneous polymerization of the historesin.
10. Handle with care, because if you touch the glass razor blade with any material that is not a brush, it will not work properly anymore. The blades can be stored in boxes with a plasticine base to prevent damage.
11. Microtome RM2265 allows manual sectioning that is best not to lose samples.
12. PI was detected with a 570- to 670-nm band pass filter. We used an FV 1000 confocal microscope (Olympus) and processed the different Z stacks with the Olympus FV software and assembled them with Photoshop CS (Adobe Systems).

Acknowledgments

We acknowledge the financial support from the Spanish Ministry of Economy and Competitiveness, through the "Severo Ochoa Programme for Centres of Excellence in R&D" 2016–2019 (SEV-2015-0533)." N.F. is indebted to the "Fundación Renta Corporation" charity and F.L-E and D.B-E. are funded by a BIO2013-43873 grant and a "Retos Colaboración" project (RTC-2014-1916-2) from the Spanish Ministry of Economy and Competitiveness, respectively. A.I.C.-D. is the recipient of a European Research Council, ERC Consolidator Grant (ERC-2015-CoG - 683163).

References

1. Aloni R, Griffith M (1991) Functional xylem anatomy in root-shoot junctions of six cereal species. Planta 184:123–129
2. Hochholdinger F, Park WJ, Sauer M et al (2004) From weeds to crops: genetic analysis of root development in cereals. Trends Plant Sci 9:42–48
3. Paterson AH, Bowers JE, Bruggmann R et al (2009) The *Sorghum bicolor* genome and the diversification of grasses. Nature 457:551–556
4. Price HJ, Dillon SL, Hodnett G et al (2005) Genome evolution in the genus *Sorghum* (Poaceae). Ann Bot 95:219–227
5. Swigoňová Z, Lai J, Ma J, Ramakrishna W, Llaca V, Bennetzen JL, Messing J (2004) Close split of sorghum and maize genome progenitors. Genome Res 14:1916–1923
6. Howe A, Sato S, Dweikat I et al (2006) Rapid and reproducible *Agrobacterium*-mediated transformation of sorghum. Plant Cell Rep 25:784–791
7. Blomstedt CK, Gleadow RM, O'Donnell N et al (2012) A combined biochemical screen and TILLING approach identifies mutations in *Sorghum bicolor* L. Moench resulting in acyanogenic forage production. Plant Biotechnol J 10:54–66
8. Nida H, Blum S, Zielinski D, Srivastava DA, Elbaum R, Xin Z, Erlich Y, Fridman E, Shental N (2016) Highly efficient *de novo* mutant identification in a *Sorghum bicolor* TILLING population using the ComSeq approach. Plant J 86:349–359
9. Xin Z, Wang ML, Barkley NA et al (2008) Applying genotyping (TILLING) and phenotyping analyses to elucidate gene function in a chemically induced sorghum mutant population. BMC Plant Biol 8:103
10. Foley JA, Ramankutty N, Brauman KA et al (2011) Solutions for a cultivated planet. Nature 478:337–342
11. Jenks MA, Joly RJ, Peters PJ et al (1994) Chemically induced cuticle mutation affecting epidermal conductance to water vapor and disease susceptibility in *Sorghum bicolor* (L.) Moench. Plant Physiol 105:1239–1245
12. Jordan WR, Shouse PJ, Blum A et al (1984) Environmental physiology of sorghum. II. Epicuticular wax load and cuticular transpiration. Crop Sci 24:1168–1173
13. Zhang J, Kirkham MB (1996) Antioxidant responses to drought in sunflower and sorghum seedlings. New Phytol 132:361–373
14. Borrell AK, Hammer GL, Henzell RG (2000) Does maintaining green leaf area in sorghum improve yield under drought? II. Dry matter production and yield. Crop Sci 40:1037–1048
15. Borrell AK, Mullet JE, George-Jaeggli B, van Oosterom EJ, Hammer GL, Klein PE, Jordan DR (2014) Drought adaptation of stay-green sorghum is associated with canopy development, leaf anatomy, root growth, and water uptake. J Exp Bot 65:6251–6263
16. Burow GB, Franks CD, Acosta-Martinez V et al (2008) Molecular mapping and characterization of *BLMC*, a locus for profuse wax (bloom) and enhanced cuticular features of Sorghum (*Sorghum bicolor* (L.) Moench.). Theor Appl Genet 118:423–431
17. Kebede H, Subudhi PK, Rosenow DT et al (2001) Quantitative trait loci influencing drought tolerance in grain sorghum (*Sorghum bicolor* L. Moench). Theor Appl Genet 103:266–276
18. Sanchez AC, Subudhi PK, Rosenow DT et al (2002) Mapping QTLs associated with drought resistance in sorghum (*Sorghum bicolor* L. Moench). Plant Mol Biol 48:713–726
19. Li R, Han Y, Lv P et al (2014) Molecular mapping of the brace root traits in sorghum (*Sorghum bicolor* L. Moench). Breed Sci 64:193–198

20. Hetz W, Hochholdinger F, Schwall M et al (1996) Isolation and characterization of *rtcs*, a maize mutant deficient in the formation of nodal roots. Plant J 10:845–857
21. Salih AA, Ali IA, Lux A et al (1999) Rooting, water uptake, and xylem structure adaptation to drought of two sorghum cultivars. Crop Sci 39:168–173
22. Mace ES, Singh V, Van Oosterom EJ et al (2011) QTL for nodal root angle in sorghum (*Sorghum bicolor* L. Moench) co-locate with QTL for traits associated with drought adaptation. Theor Appl Genet 124:97–109
23. Rajkumar, Fakrudin B, Kavil SP et al (2013) Molecular mapping of genomic regions harbouring QTLs for root and yield traits in sorghum (*Sorghum bicolor* L. Moench). Physiol Mol Biol Plants 19:409–419
24. Fàbregas N, Li N, Boeren S et al (2013) The BRASSINOSTEROID INSENSITIVE1-LIKE3 signalosome complex regulates *Arabidopsis* root development. Plant Cell 25:3377–3388
25. González-García MP, Vilarrasa-Blasi J, Zhiponova M et al (2011) Brassinosteroids control meristem size by promoting cell cycle progression in *Arabidopsis* roots. Development 138:849–859
26. Vilarrasa-Blasi J, González-García MP, Frigola D et al (2014) Regulation of plant stem cell quiescence by a brassinosteroid signaling module. Dev Cell 30:36–47
27. Kinoshita T, Caño-Delgado A, Seto H et al (2005) Binding of brassinosteroids to the extracellular domain of plant receptor kinase BRI1. Nature 433:167–171
28. Mantilla Perez MB, Zhao J, Yin Y et al (2014) Association mapping of brassinosteroid candidate genes and plant architecture in a diverse panel of *Sorghum bicolor*. Theor Appl Genet 127:2645–2662
29. Kir G, Ye H, Nelissen H et al (2015) RNA interference knockdown of BRASSINOSTEROID INSENSITIVE1 in maize reveals novel functions for brassinosteroid signaling in controlling plant architecture. Plant Physiol 169:826–839
30. Truernit E, Bauby H, Dubreucq B et al (2008) High-resolution whole-mount imaging of three-dimensional tissue organization and gene expression enables the study of phloem development and structure in *Arabidopsis*. Plant Cell 20:1494–1503

Chapter 16

Brassinosteroid Action in Plant Abiotic Stress Tolerance

Priti Krishna, Bishun D. Prasad, and Tawhidur Rahman

Abstract

Brassinosteroids (BRs) are a class of plant steroidal hormones that play essential roles in plant growth and development. Systematic studies had first been undertaken concomitantly to determine both the effects of exogenous BR on stress phenotypes of *Arabidopsis thaliana* and *Brassica napus* (rapeseed) seedlings and the expression of stress marker genes in BR-treated and untreated seedlings. When reproducible and convincing evidence of the role of BR in stress tolerance had been obtained, molecular mechanisms underlying the ability of BR to confer tolerance against heat, cold, drought, and salt stress, as well as pathogen resistance were studied with several molecular approaches and tools. The results of these studies have together provided valuable insights into how BRs, through their control of many basic cellular processes and stress responses, promote vigor in plants and prepare the plant to mount a dynamic response upon environmental challenges. Protocols to assess BR effects on abiotic stress tolerance in *Arabidopsis* and rapeseed seedlings are described here and they can be fine-tuned and adapted for other plant species.

Key words Brassinosteroid, 24-Epibrassinolide, Heat stress, Cold stress, Drought stress, Salt stress, *Arabidopsis*, *Brassica napus*

1 Introduction

Brassinosteroids (BRs) are plant steroid hormones, which play essential roles in plant growth and developmental programs [1, 2] The BR signal is perceived by the membrane-localized receptor BRASSINOSTEROID INSENSITIVE1 (BRI1) that, upon BR binding, dimerizes with the coreceptor BRI1-ASSOCIATED RECEPTOR KINASE1 (BAK1). The kinase domains of BRI1 and BAK1 undergo sequential transphosphorylation, thereby turning on the signaling reactions. BR signaling activated by BRI1 inactivates the negative regulator BRASSINOSTEROID INSENSITIVE2 (BIN2) that, in turn, triggers activation of two closely related transcription factors, BRI1-ETHYL METHANESULFONATE-SUPPRESSOR1 (BES1) and BRASSINAZOLE-RESISTANT1 (BZR1), and expression of their target genes [3, 4].

In addition to their essential roles in plant growth and development, BRs confer tolerance to a range of abiotic stresses [5, 6].

Eugenia Russinova and Ana I. Caño-Delgado (eds.), *Brassinosteroids: Methods and Protocols*, Methods in Molecular Biology, vol. 1564, DOI 10.1007/978-1-4939-6813-8_16,

Preliminary studies conducted in the 1980s and early 1990s to assess the BR ability to promote stress tolerance led to inconsistent results, probably resulting from differing BR application methods (seed soak, root soak, or foliar spray), as well as different developmental stages at which BR was applied [5]. We tested different modes of BR application for reproducibility of BR effects on plant stress tolerance and found that feeding BR via roots for 2–3 weeks prior to stress exposure produced the most reproducible results. As BR is a major growth hormone, a short-term treatment followed by immediate exposure to stress may simplistically translate into tricking a plant that has been induced for BR-mediated growth responses, such as cell elongation, to cope with the imposed stress. Thus, whereas short-term BR treatment may provide useful information on immediate gene expression and physiological BR-mediated changes, BR effects on stress tolerance will probably not be fully recognized. By contrast, long-term BR treatment may lead to changes in meristem size, xylem patterning, root length, and other traits that may also contribute to stress tolerance.

BR effects on basic thermotolerance were tested by growing *Brassica napus* (rapeseed) and *Arabidopsis thaliana* seedlings on Murashige and Skoog (MS) medium [7], containing either 1 μM 24-epibrassinolide (EBR) or an equivalent volume of ethanol (EBR solvent) for 2 or 3 weeks and then subjecting the seedlings to heat stress. EBR-treated seedlings had higher survival rates than untreated seedlings, concomitant with higher accumulation of heat shock proteins (HSPs), resulting from enhanced protein synthesis in these seedlings [8–10], and with increased levels of the major plant stress hormone abscisic acid (ABA) [11].

Studies of BR effects on chilling and freezing stress include treatment of cell suspension cultures [12] and spraying of seedlings before exposure to cold [13, 14], as well as dipping fruits in a BR solution before cold storage [15]. Genetic evidence for the role of BRs role in cold stress tolerance originates from transgenic *Arabidopsis* seeds expressing the BR-biosynthetic gene *DWARF4* under the control of a seed-specific oleosin promoter [16]. Cold-induced expression of most *BZR1* orthologs in *Brassica rapa* (turnip) also supports a role for BR in cold stress tolerance [17].

Positive effects of BR on drought stress tolerance have been demonstrated in *Arabidopsis* and rapeseed, both at the phenotypic and gene expression levels [18, 19]. In line with these observations, the semidwarf allelic mutants of *Hordeum vulgare* (barley) with impaired activity of the BR-biosynthetic enzyme C6-oxidase are compromised in drought tolerance, and, in comparison to the wild type, accumulate less osmoprotectants and have a lower net photosynthetic rate in response to drought stress [20].

The seed germination phase in a plant life cycle is highly sensitive to salt. BR enhanced the germination percentages in *Arabidopsis* and rapeseed [18]. In seedlings sprayed with BR, the H_2O_2 content, electrolyte leakage, and oxidative damage in response to salt stress were reduced when compared to controls [14, 21]. *Oryza sativa* (rice) seeds soaked in EBR solution exhibited better growth and more efficient stress responses than untreated control seedlings [22]. That BR signaling is involved in salt tolerance is evidenced by the greater salt sensitivity of *bri1-9* [23] and the *br signaling kinase5* (*bsk5*) [24] mutant seedlings *vs* the wild type.

In summary, a role for BR in abiotic stress tolerance is now firmly entrenched in the literature and the underlying mechanisms are beginning to be uncovered. There is considerable evidence that BR and ABA collaborate in developing stress tolerance [11]. Signaling mediated by secondary messengers, such as H_2O_2 [14], NO [21], and Ca^{2+} [25], by mitogen-activated protein kinases (MAPK) [26], and by the BR transcription factors BES1/BZR1 [17, 23, 24] has all been implicated in BR-mediated abiotic stress tolerance. Future studies should incorporate the big picture of the BR role in multiple plant processes when assessing its function in stress tolerance, i.e., how BR-mediated architectural, anatomical, and metabolic changes in response to stress combine with the known specific stress responses to provide an overarching advantage to the plant for overcoming the negative impacts of different stresses.

2 Materials

2.1 Plant Material

1. *Arabidopsis thaliana* (L.) Heynh. (accession Columbia-0 [Col-0]).
2. *Brassica napus* cv. Westar (rapeseed).

2.2 Solutions and Equipment for Seed Sterilization

1. 75 % (v/v) ethanol (dilute 95 % [v/v] ethanol with deionized, sterile water to achieve 75 % ethanol).
2. 1.05 % (v/v) sodium hypochlorite; 20 % (v/v) bleach with or without 0.1 % (v/v) polyoxyethylenesorbitan monolaurate (Tween 20) can be used as well.
3. Autoclaved distilled water.
4. Balance.
5. Sterile 1.5-mL Eppendorf tubes or 50-mL tubes (Falcon conical centrifuge tubes).
6. Autoclave.
7. Laminar flow hood.
8. Sterile micropipette tips (0.1–10 and 10–200 μL) to be opened only in the hood.
9. Waste container.
10. Sterilized filter paper.

2.3 Solutions and Equipment for Thermotolerance Assay

1. Solid growth medium: prepare 1× Murashige and Skoog (1× MS) medium supplemented with 1 % (w/v) sucrose, adjust pH to 5.7–5.8 with 1 M KOH, add phyto agar to 1 % (w/v?) (or 0.8 %), and sterilize by autoclaving at 121 °C for 20 min.
2. 24-epibrassinolide (EBR): prepare a 10 mM stock solution by adding absolute ethanol directly to the bottle containing the EBR powder, mix, and store the stock solution at −20 °C.
3. Ethanol (absolute alcohol, nondenatured) as EBR solvent.
4. Balance.
5. pH meter.
6. Autoclave.
7. Sterile measuring cylinders.
8. Magenta GA-7 vessel.
9. Petri plates (150×15 mm).
10. Sterile forceps and tweezers.
11. Parafilm.
12. Plant growth chamber (such as Conviron TC16).
13. Incubator with temperature ranging from ambient to 60 °C.

2.4 Solutions and Equipment for Cold Stress Tolerance Assay

1. Materials listed in Subheading 2.3.
2. Low temperature incubator.

2.5 Solutions and Equipment for Drought Tolerance Assay

1. Materials listed in Subheading 2.3.
2. Coarse sand (*Arabidopsis*) or soil (Pro-Mix BX for *B. napus* or other large plants).
3. Plastic pots for growing plants.

2.6 Solutions and Equipment for Salt Tolerance Assay

1. Materials listed in Subheading 2.3.
2. NaCl.
3. Sterile Petri plates (100×15 mm for germination studies and 120×120 mm for root length assay).
4. Soil (Pro-Mix BX) and plastic pots for growing plants.

3 Methods

3.1 Sterilization of Seeds

1. Take the desired amount of seeds in 1.5-mL microcentrifuge tubes (*Arabidopsis*), or 50-mL conical centrifuge tubes (rapeseed).
2. Add 75 % (v/v) ethanol and incubate for 1 min by inverting or rotating the tubes several times.

3. Remove ethanol and wash seeds 4–5 times in sterile-distilled water.
4. Add 1.05 % (v/v) sodium hypochlorite and incubate for 20 min with inverting tube several times.
5. Remove bleach as much as possible and rinse 4–5 times with sterile distilled water [8, 9] (*see* **Note 1**).
6. Transfer seeds to sterile filter paper and partially air-dry under laminar flow (optional).

3.2 Thermotolerance Assay in Arabidopsis

1. Prepare 1× MS agar medium, sterilize. When the medium has cooled to approximately 45 °C, add EBR to a final concentration of 1 μM or other desired concentrations (*see* **Note 3**) by using the 10 mM stock solution. To maintain an accurate hormone concentration, tip the flask and add the hormone directly into the medium as opposed to the side of the flask/bottle. Add the same volume of ethanol (EBR solvent) to the medium used for growing the control (untreated) seedlings. Mix and pour an equal amount of medium into Petri plates and allow the medium to solidify.
2. Place sterilized seeds on the surface of the medium with sterilized tweezers or a sterile pipette tip. Place the lid and seal plates with parafilm.
3. Incubate plates for 3 days in the dark at 4 °C to encourage synchronized germination.
4. Incubate plates for 21 days in a plant growth chamber (TC16) at 22 °C under a 16-h light/8-h dark photoperiod and a light intensity of 80 μE $m^{-2}s^{-1}$ (*see* **Notes 1** and **2**).
5. Expose 21-day-old seedlings to 43 °C for varying lengths of time (1, 2, 3, 4, or 5 h), and then return plates to 22 °C to allow seedling recovery for 7 days (Table 1).
6. Score seedlings as alive, dead, or recovering [9]. Seedlings exposed to 43 °C for 3 and 4 h show a clear EBR effect (Fig. 1) (*see* **Notes 4** and **6**).

3.3 Thermotolerance Assay in Rapeseed

1. Prepare medium as in Subheading 3.2 (**step 1**). Pour equal amounts of the medium in sterilized magenta boxes and let them solidify (*see* **Note 3**).
2. Place nine sterilized rapeseed seeds on the solidified medium with a long sterilized forceps and incubate at 4 °C for 3 days in the dark to encourage synchronized germination.
3. Transfer magenta boxes to a plant growth chamber maintained at 20 °C and 16-h light/8-h dark photoperiod with a light intensity of 125 μE m^{-2} s^{-1} (*see* **Note 2**).
4. Expose 14-day-old seedlings to 45 °C for 2, 3, 4, or 5 h and return magenta boxes to 20 °C for seedling recovery for 7 days (Table 1).

Table 1
Assays for abiotic stress tolerance

Plant species	Seedling age	Stress treatment	Time of scoring seedling phenotype	Phenotype score
Thermotolerance assay				
Arabidopsis	Grown on MS agar medium ± EBR for 21 days	43 °C for 1, 2, 3, 4, or 5 h	After 7 days at 22 °C	Alive, dead, or recovering
Brassica napus	Grown on MS agar medium ± EBR for 14 days	45 °C for 2, 3, 4, or 5 h	After 7 days at 20 °C	Alive, dead, or recovering
Drought tolerance assay				
Arabidopsis	Grown on MS agar medium ± EBR for 21 days and then on sand for 5 days	Water withheld for up to 96 h	After rewatering for 2 days	Alive or dead
B. napus	Grown on MS agar medium ± EBR for 14 days and then on sand for 3 days	Water withheld for up to 60 h	After rewatering for 2 days	Alive or dead
Seed germination under salt stress				
Arabidopsis	Mature seed	MS agar medium ± EBR containing 0, 50, 100, 150, or 200 mM NaCl	After 24, 48, 72, and 96 h incubation at 22 °C	Radicle emergence
B. napus	Mature seed	MS agar medium ± EBR containing 0, 50, 100, 150, 200, 250, or 300 mM NaCl	After 24, 48, 72, and 96 h incubation 20 °C	Radicle emergence
Salinity tolerance assay				
Arabidopsis	Grown in pots containing Pro-Mix BX for 3 weeks	Watered with 150 mM NaCl solution on alternate days	When symptoms appear (3–5 weeks)	Alive or dead
Root length assay for salinity tolerance				
Arabidopsis	Mature seed	MS agar medium ± EBR containing 0, 50, 100, 150 or 200, mM NaCl	After 2 weeks of growth on vertical plates	Root length

5. Score seedlings as dead (bleached cotyledons and true leaves, without growth subsequent to heat stress), alive (negligible to limited bleaching of the cotyledons), or recovering (bleached cotyledons with green leaves). Seedlings exposed to 45 °C for 4 h show a clear EBR effect [8] (*see* **Notes 4** and **6**).

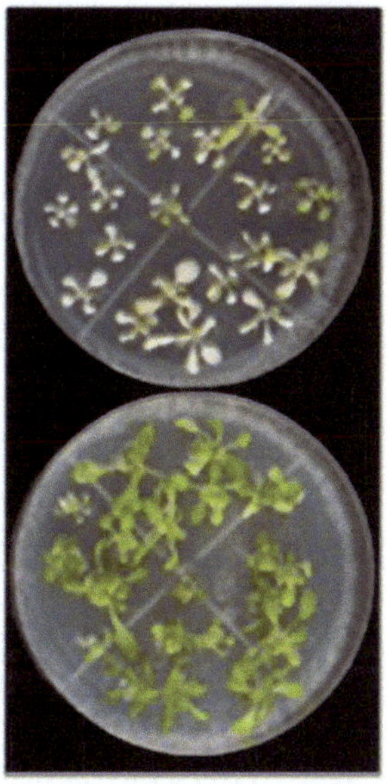

Fig. 1 BR effect on basic thermotolerance in *Arabidopsis*. *Arabidopsis thaliana* (accession Landsberg *erecta*) seedlings were grown for 21 days on MS agar medium containing either 1 μM 24-epibrassinolide (EBR) or an equivalent amount of ethanol (control). The seedlings were exposed to 43 °C for 3 h and then recovered at 22 °C for 7 days. The photograph was taken 7 days after recovery from heat stress

3.4 Drought Tolerance Assay in Arabidopsis

1. Grow seedlings for 21 days on 1× MS agar medium as described in Subheading 3.2 (**step 1**).
2. Transplant 21-day-old healthy seedlings into pots containing equal amounts of coarse sand.
3. Allow seedlings to reestablish growth for 5 days (*see* **Notes 2** and **5**)
4. Subject seedlings to drought stress by withholding water for up to 96 h (*see* **Note 3**).
5. After the drought stress, water the seedlings for the next 2 days (Table 1).
6. Score seedlings as dead or alive [18] (*see* **Notes 4** and **6**).

3.5 Drought Tolerance Assay in Rapeseed

1. Grow seedlings for 14 days on 1× MS agar medium as described in Subheading 3.3 (**step 1**-**step 3**).
2. Transplant 14-day-old healthy seedlings into pots containing equal amounts of coarse sand.
3. Allow seedlings to reestablish growth for 3 days (*see* **Notes 2** and **5**).
4. Subject seedlings to drought stress by withholding water for up to 60 h (*see* **Note 3**).

5. After the drought stress, water the seedlings for the next 2 days (Table 1).
6. Score seedlings as dead or alive [18] (*see* **Notes 4** and **6**).

3.6 Germination Assay on Salt

1. Prepare 1× MS agar medium plates containing 0, 50, 100, 150, or 200 mM NaCl (*Arabidopsis*) or 0, 50, 100, 150, 200, 250, or 300 mM NaCl (rapeseed) and either 1 μM EBR or an equal amount of ethanol (*see* **Note 3**).
2. Place sterilized seeds on the surface of the medium and incubate plates as described in Subheadings 3.2 and 3.3. Use Petri plates instead of magenta boxes, depending on the plant species (*see* **Note 2**).
3. Score germination at 24, 48, 72, and 96 h [18] or seedling survival after 15 days [11]. Score seeds as germinated when the radicle has emerged from the seed coat (Table 1) (*see* **Note 4**).

3.7 Salinity Tolerance Assay in Soil

This assay is described for BR-biosynthetic and signaling mutants or overexpression lines in which no BR treatment is required. For experiments involving BR treatment, follow Subheadings 3.2 and 3.3 for growth on EBR and then transplant seedlings in pots filled with Pro-Mix BX. After 5 days of seedling recovery from the transplantation shock (*see* **Note 5**), follow salt treatment as described.

1. Place surface-sterilized, equal-sized *Arabidopsis* seeds in plastic pots (8 × 7.5 cm^2) containing the same amount of Pro-Mix BX per pot.
2. Grow seedlings in a growth chamber for 3 weeks with watering every other day (*see* **Notes 2** and **5**).
3. Expose 3-week-old seedlings to salt stress by watering with either 150 mM NaCl solution or with an equal volume of water (control) every other day for 3–5 weeks until symptoms appear (Table 1) (*see* **Note 3**).
4. Score seedlings as living or dead after the stress treatment or after recovery from stress. Seedlings with intact green leaves in the middle of the rosette that continue to grow are scored as survivors (*see* **Notes 4** and **6**).

3.8 Root Length Assay for Salinity Tolerance

1. Prepare 1× MS medium plates (square) containing 0, 50, 100, 150, or 200 mM NaCl and either 1 μM EBR or an equal volume of ethanol (control) (*see* **Note 3**).
2. Place sterilized *Arabidopsis* seeds in a row on the surface of the medium in Petri plates. Seal plates and place them vertically in the growth chamber so that roots grow on the surface of the medium as opposed to growing into the medium (*see* **Note 2**).
3. Measure root length after 2 weeks (Table 1) (*see* **Note 4**).

4 Notes

1. Detailed information on growing *Arabidopsis* can be found at the Arabidopsis Biological Resource Center site (https://abrc.osu.edu/seed-handling).
2. Petri plates, magenta boxes, and pots containing seeds or seedlings should be randomly moved around in the growth chamber to avoid position effects.
3. For optimizing BR concentration and stress conditions, initially use a range of concentrations (e.g., 0.1, 0.5, 1.0. 1.5, and 2.0 μM EBR) and a range of stress conditions (e.g., expose plants to 35 °C, 38 °C, 40 °C, and 45 °C). Most hormones, including BRs, are inhibitory at high concentrations. Furthermore, an optimal BR concentration may vary with plant species, plant cells, tissues, and organs to which BRs are applied, the mode of application, and the BR type. The length of the BR treatment is a very important factor in assessing BR effects. We found that growing seedlings on BR for 2–3 weeks produced clear and reproducible effects on stress phenotypes of seedlings.
4. All experiments should have an appropriate number of replicates and should be repeated three times for reproducibility.
5. When growing seedlings in pots, add an equal amount of soil or sand to all pots and water pots with an equal volume of water. It is crucial that the soil water capacity is maintained the same for all plants.
6. Methods described here are for testing BR effects on seedling phenotypes after stress. Concomitant with stress phenotypes, gene expression changes [8–11], and some physiological characteristics, such as thiobarbituric acid reactive substances as an indicator of oxidative stress [9], chlorophyll fluorescence as an indicator of photosystem II activity, and changes in other hormone levels [11] were also studied. Other studies have mainly focused on mapping the BR impact on physiological stress-induced changes. In designing experiments to study BR effects, it is important to keep in mind that the results are tied in, to some extent, with mode, timing, and length of BR application.

Acknowledgements

We thank the Natural Sciences and Engineering Research Council of Canada for supporting the studies and research personnel. B.D.P. was supported through a Postdoctoral Doctoral Fellowship, and T.R. through Ph.D. Research and Teaching Assistantships.

References

1. Clouse SD (2011) Brassinosteroids. The Arabidopsis Book 9:e0151
2. Yang C-J, Zhang C, Lu Y-N et al (2011) The mechanisms of brassinosteroids' action: from signal transduction to plant development. Mol Plant 4:588–600
3. Zhu J-Y, Sae-Seaw J, Wang Z-Y (2013) Brassinosteroid signalling. Development 140: 1615–1620
4. Belkhadir Y, Jaillais Y (2015) The molecular circuitry of brassinosteroid signalling. New Phytol 206:522–540
5. Krishna P (2003) Brassinosteroid-mediated stress responses. J Plant Growth Regul 22:289–297
6. Divi UK, Krishna P (2010) Brassinosteroids confer stress tolerance. In: Hirt H (ed) Plant stress biology: from genomics to system biology. Wiley, New York, pp 119–135
7. Murashige T, Skoog F (1962) A revised medium for rapid growth and bio assay with tobacco tissue cultures. Physiol Plant 15:473–497
8. Dhaubhadel S, Chaudhary S, Dobinson KF et al (1999) Treatment with 24-epibrassinolide, a brassinosteroid, increases the basic thermotolerance of *Brassica napus* and tomato seedlings. Plant Mol Biol 40:333–342
9. Divi UK, Rahman T, Krishna P (2010) Brassinosteroid-mediated stress tolerance in Arabidopsis shows interactions with abscisic acid, ethylene and salicylic acid pathways. BMC Plant Biol 10:151
10. Dhaubhadel S, Browning KS, Gallie DR et al (2002) Brassinosteroid functions to protect the translational machinery and heat-shock protein synthesis following thermal stress. Plant J 29:681–691
11. Divi UK, Rahman T, Krishna P (2016) Gene expression and functional analyses in brassinosteroid-mediated stress tolerance. Plant Biotechnol J 14:419–432
12. Liu Y, Jiang H, Zhao Z et al (2011) Abscisic acid is involved in brassinosteroids-induced chilling tolerance in the suspension cultured cells from *Chorispora bungeana*. J Plant Physiol 168:853–862
13. Deng X-G, Zhu T, Zhang D-W et al (2015) The alternative respiratory pathway is involved in brassinosteroid-induced environmental stress tolerance in *Nicotiana benthamiana*. J Exp Bot 66:6219–6232
14. Wei L-J, Deng X-G, Zhu T et al (2015) Ethylene is involved in brassinosteroids induced alternative respiratory pathway in cucumber (*Cucumis sativus* L.) seedlings response to abiotic stress. Front Plant Sci 6:982
15. Gao H, Kang L, Liu Q et al (2015) Effect of 24-epibrassinolide treatment on the metabolism of eggplant fruits in relation to development of pulp browning under chilling stress. J Food Sci Technol 52:3394–3401
16. Divi UK, Krishna P (2010) Overexpression of the brassinosteroid biosynthetic gene *AtDWF4* in *Arabidopsis* seeds overcomes abscisic acid-induced inhibition of germination and increases cold tolerance in transgenic seedlings. J Plant Growth Regul 29:385–393
17. Saha G, Park J-I, Jung H-J et al (2015) Molecular characterization of BZR transcription factor family and abiotic stress induced expression profiling in *Brassica rapa*. Plant Physiol Biochem 92:92–104
18. Kagale S, Divi UK, Krochko J et al (2007) Brassinosteroid confers tolerance i*n Arabidopsis thaliana* and *Brassica napus* to a range of abiotic stresses. Planta 225:353–364
19. Sahni S, Prasad BD, Liu Q et al (2016) Overexpression of the brassinosteroid biosynthetic gene *DWF4* in *Brassica napus* simultaneously increases seed yield and stress tolerance. Sci Rep 6:28298
20. Janeczko A, Gruszka D, Pociecha E et al (2015) Physiological and biochemical characterisation of watered and drought-stressed barley mutants in the *HvDWARF* gene encoding C6-oxidase involved in brassinosteroid biosynthesis. Plant Physiol Biochem 99:126–141
21. Zhu T, Deng X-G, Tan W-R et al (2015) Nitric oxide is involved in brassinosteroid-induced alternative respiratory pathway in *Nicotiana benthamiana* seedlings' response to salt stress. Physiol Plant 156:150–163
22. Sharma I, Ching E, Saini S et al (2013) Exogenous application of brassinosteroid offers tolerance to salinity by altering stress responses in rice variety Pusa Basmati-1. Plant Physiol Biochem 69:17–26
23. Cui F, Liu L, Zhao Q et al (2012) *Arabidopsis* ubiquitin conjugase UBC32 is an ERAD component that functions in brassinosteroid-mediated salt stress tolerance. Plant Cell 24:233–244
24. Li Z-Y, Xu Z-S, He G-Y et al (2012) A mutation in *Arabidopsis BSK5* encoding a brassinosteroid-signalling kinase protein affects responses to salinity and abscisic acid. Biochem Biophys Res Commun 426:522–527
25. Yan J, Guan L, Sun Y et al (2015) Calcium and ZmCCaMK are involved in brassinosteroid-induced antioxidant defense in maize leaves. Plant Cell Physiol 56:883–896
26. Zhang A et al (2010) ZmMPK5 is required for the NADPH oxidase-mediated self-propagation of apoplastic H_2O_2 in brassinosteroid-induced antioxidant defence in leaves of maize. J Exp Bot 61(15):4399–4411

Index

Eugenia Russinova and Ana I. Caño-Delgado (eds.), *Brassinosteroids: Methods and Protocols*, Methods in Molecular Biology, vol. 1564, DOI 10.1007/978-1-4939-6813-8, © Springer Science+Business Media LLC 2017

FSC
www.fsc.org
MIX
Papier aus verantwortungsvollen Quellen
Paper from responsible sources
FSC® C105338

If you have any concerns about our products,
you can contact us on
ProductSafety@springernature.com

In case Publisher is established outside the EU,
the EU authorized representative is:
Springer Nature Customer Service Center GmbH
Europaplatz 3, 69115 Heidelberg, Germany

Printed by Libri Plureos GmbH
in Hamburg, Germany